2019 Amateur Radio Emergency Communications Conference

Gainesville, Florida
February 2 & 3, 2019

Gordon L. Gibby KX4Z, Editor
Joe Bassett W1WCN
Leland Gallup AA3YB
Karl Martin KG4HBN
Marvin Corbin, Florida Baptist Disaster Relief
Scott Roberts KK4ECR
Jeff Capehart W4UFL

and many others who coached us to get here!

ISBN: 9781791865948

DEDICATION

This text is dedicated to all the hard working ham radio operators who have made emergency amateur radio communications grow in skills and assets in Florida, and to all those who will follow in their footsteps.

CONTENTS

Acknowledgments   i

1   Volunteer ham radio team building that maximizes all volunteers   1
         Joe Bassett, Jr. W1WCN

2   Using ICS Documents to manage large deployments & intro to the full scale exercise   3
         Gordon Gibby KX4Z

3   How to plan / create / carry out full scale exercises and build your group   9
         Leland Gallup AA3YB

4   Lessons from Cascadia Rising and Oregon 2018 SET   23
         Gordon Gibby KX4Z

5   Section Emergency Coordinator's Message   33
         Karl Martin KG4HBN

6   Baptist Disaster Relief Services for Amateur Radio Volunteers   37
         Marvin Corbin

7   WIFI shelter bulletin system to keep shelter residents informed   41
         Gordon Gibby KX4Z

8   Introduction to Publicity / PIO   47
         Scott Roberts KK4ECR

9   Moving Traffic and Training Volunteers in ARES Nets   51
         Joe Bassett, Jr. W1WCN

10   Computer And Internet Tips for EMCOMM   57
         Jeff Capehart W4UFL

11   Wiring your radio for Signalink / digital   67
         Gordon Gibby KX4Z

12   Repeater Controller – ICS-CTRL   71
         Gordon Gibby KX4Z

13   Hands on Solar Power   79
         Gordon Gibby KX4Z

14   Emergency VHF Antenna Building   83
         Gordon Gibby KX4Z

15   Tuning a duplexer with a $110 Chinese Spectrum Analyzer   87
         Gordon Gibby KX4Z

16   Soundcard Isolator   97
         Gordon Gibby KX4Z

17   Creating Winlink Account without Internet   105
         Gordon Gibby KX4Z

ACKNOWLEDGMENTS

I'd like to acknowledge all the people who have written chapters, taken our courses, passed amateur radio exams, built stations, learned systems, and worked so hard to improve amateur radio emergency communications in northern Florida. And also the people who taught me how to get books published on Amazon – and how to create printed circuit boards that could be manufactured so inexpensively in China. And especially all the people who put up with our mistakes along the way! None of us will ever forget the 12+ hours it took us to hand wire the first set of boards!

1 VOLUNTEER HAM RADIO TEAM BUILDING THAT MAXIMIZES ALL VOLUNTEERS

by Joe Bassett, Jr W1WCN

First, let's lose the term "volunteer." The success of any cause is owed to those who are committed to that cause. Both of the World Wars were won by nations who rallied their citizens to the cause of freedom. Even in today, the overwhelming superiority of the United States' "volunteer" military is rooted in a call to excellence, honor, duty, and service inherent in preserving liberty. The tip of today's military sword is comprised of individuals who have pursue excellence beyond that of the rank and file. In a word, they answered a call.

So, let's leave behind the concept of "volunteers" and recognize the role of the call to service above self. In one sense a person who offers their service, effort, and resources has volunteered, but those who commit to the vision of an organization respond to a call. The volunteer comes and goes, wavers in adversity, harbors the possibility of withdrawing from the effort. Those that commit to a cause, those that answer a call, are compelled to duty by the intrinsic value of a higher vision.

This is not to confuse the difference between professionals and amateurs. History is filled with amateurs who were committed to a cause the overshadowed professionals. The distinction between professional and amateur comes from compensation, not ability. Professional golfers are still measured against the great golfer of the early 1900s, Bobby Jones, yet Mr. Jones competed as an amateur, maintaining his standing as a renowned attorney.

Professionals built the Titanic, amateurs built the arc.

Also, answering a call to service is not contingent upon ability. Today's culture tends to emphasize intelligence above effort. However, the theme of hard work resounds throughout stories of corporate, athletic, cultural, and academic success. Sure enough, great thought and insight plays a role in success, but great success can be experienced with even a moderate IQ. Rarely does success come from genius alone. In the end, smart always comes from hard work.

In short, members of a successful team don't need to be superstars, they don't need to come to the organization pre-packaged. In fact, pre-packaged superstars are often a hindrance to a team's success. Successful teams are comprised of ordinary people who share a common call to serve and are led well.

John Maxwell opens his book "21 Irrefutable Laws of Leadership" with The Law of the Lid. Leadership ability – for better or for worse – determines the effectiveness and impact of an organization. Obviously, the onus for team success rests squarely on the shoulders of leadership and at any moment any member should be prepared to lead in some manner. That particular leader might be at the top of the organizational chart, or they might be a subordinate who fulfills a leadership role, but the ability of organizational leadership is the lid to the organization's success.

It's commonly understood that management and leadership are distinct skillsets. It is less common to understand the distinction. In short, management sees personnel as a commodity, a resource to be exploited for the profit of stakeholders. Leadership envisions a bright future, values people for their own sake, understands the contribution those people can make, and inspires the passion to improve themselves, their colleagues, the organization and society.

Patrick Lencioni's "The Five Dysfunctions of a Team" is an excellent resource for leaders who seek to improve their ability to lead well, not only for their own benefit, but for the benefit of everyone in the organization and those whom the organization serves. These dysfunctions are: The Absence of Trust, Fear of Conflict, Lack of Commitment, Avoidance of Accountability, Inattention to Results.

The base of the Dysfunction Pyramid is an Absence of Trust. This trust is multi-faceted: the team's trust of the leader, the leader's trust of the team members, the members' trust of each other, and everyone's trust in themselves. The heart of this trust is to know one's self and those around them. There are myriad tools that can build this understanding and trust, such as the Myers-Briggs Personality survey or Structural Enneagram. Of course, simply spending time together in various settings and circumstances is beneficial for building trust.

Regardless of how the absence of trust is overcome, the elimination of this first dysfunction is paramount for maximizing the team's efforts, the members' personal growth, and the benefit to society. In turn, overcoming all five dysfunctions doesn't guarantee organizational success, but it does provide an environment for thriving.

2 USING ICS DOCUMENTS TO MANAGE LARGE DEPLOYMENTS
&
INTRO TO THE FULL SCALE EXERCISE

by Gordon Gibby KX4Z

At its core, the Incident Command System is all about maintaining some semblance of ORDER in the midst of a Response to a disaster, which is inherently chaotic. The paperwork in the ICS system seems to be calculated to give you a place to write down the things that everyone needs to know, and keep straight.

For the Ham radio operator, there seem to be a FEW KEY DOCUMENTS with which you just have to be familiar.

ICS-205
The ICS-205 tells you the Frequencies, offsets, tones (for FM repeaters) and MODES of communications in your area. It may include a Command Net frequency, possibly some alternates should the repeater or frequency become unusable, and for Hams it may also include some hf single sideband phone frequencies, possibly even some digital frequencies for packet communications on VHF or WINLINK or other mode on HF. If things are really well organized it may also give you some insight into what other groups are using for their communications – and that might allow you to listen in and gain more insight into what is happening all around you! Obviously, don't transmit on any frequency you're not supposed to unless there is imminent risk to life or property and there is no other way to deal with it!!!

INCIDENT RADIO COMMUNICATIONS PLAN (ICS-205)										
1. Incident Name: EXAMPLE			**2. Date/Time Prepared:** Date: Time:			**3. Operational Period** Date From: FEB 2 Date To: FEB2 Time From: 0900 Time To:1300				
4. Basic Radio Channel Use:										
Zone Grp.	Ch #	Function	Channel Name / Trunked Radio System Talkgroup	Assign-ment	RX Freq. N or W	RX TON E / NAC	TX FREQ N or W	TX TON E / NAC	MOD E (A, D, or M)	Remarks
		Com-mand	HF NET	Ama-teur	3.900 LSB	N/A	3.900 LSB	N/A	A	adjust slightly up or down as required for interference
		Tactical	HF NET - ALTER-NATE	Ama-teur	7.260 LSB	NA	7.260 LSB	N/A	A	adjust slightly up or down as required for interference
		Tactical	K4GNV82 Repeater	Ama-teur	146.82 0 W	N/A	146.22 0 W	123	A	Primary Command Net Frequency
5. Special Instructions										

Example ICS-205

Key to our exercises is recognizing that the ICS-205 tells you *where*, out of all the millions and billions of possible frequencies, you might MEET UP with other teams if your primary systems quit!!!   So you might want to have someone LISTENING to some of those alternates, right?   Because someone might be trying to establish communications!

ICS-201

This is the simple initial BRIEFING document that might constitute the entire "incident action plan" at the beginning of a response. It may include maps, sketches, situations, cautions, assignments, organizational information, frequencies, phone numbers -- just a huge host of information.   It is pretty self-explanatory. Our group has used it to manage all kinds of operations, from holding a Technician Class to putting in new antennas on the top of the building -- it works well because it has sections for everything you need to know-- WHO, WHAT, WHERE, WHEN .... the obvious stuff. Anybody picking one up figures it out pretty quickly!

 It is likely that as the response becomes more orderly, this document is going to be superseded in subsequent work periods with SEVERAL ics documents which allow for more amplification of situational awareness information, including

     ICS 202   Incident Objectives
     ICS 204   Incident Organization (who is in charge of what)
     ICS 205   Radio Communications
     ICS 206   Medical

ICS 208   Safety Plan

Blank form versions of all of these, in popular word processor formats are readily available, as are PDF forms that you can fill in and print.   I have also created blank versions that I can insert into texts easily.

Word Processor Forms: https://www.fda.gov/EmergencyPreparedness/NIMS/ucm268797.htm
PDF: https://training.fema.gov/icsresource/icsforms.aspx

ICS-205A
The ICS-205A is used by several groups to tell how to reach each group -- in other words, it tells you WHAT THEY ARE SUPPOSED TO BE LISTENING TO.   There is a section at the bottom of the ICS204 that does the same thing.   Use whichever form suits your purpose better.

ICS-205A

1. Incident Name: **EXERCISE** **villa nova**	2. DATE / TIME PREPARED: Date: DEC 4 2018 Time: 0390	3. OPERATIONAL PERIOD Date From: Feb 2 2019 Time From: 0900 Date TO: Feb 2, 2019 Time TO: 1230
3. Basic Local Communications Information:	Local WINLINK Assets: KX4Z hf gateway operating on low power (5 watts) KX4Z-10 VHF winlink gateway operating on 2 frequencies: • 145.070 - **with digipeater/node NEWB on 145.070 south of High Springs on high tower** • 145.030 - **with digipeater/node GARC2 on 145.030 on UF Campus high dormitory**	
Incident Assigned Position	**Name**	**Method(s) of Contact (phone, pager, cell, etc.)**
Florida State COML	Gerald Busby [fictitious]	**FLCOML@WINLINK.ORG** (sole means of contact)
Amateur Comm Leader	Karl Martin KG4HBN	1. VHF Command Net (frequency per ICS-205) 2. HF Net (frequency per ICS-205) 3. WINLINK: KG4HBN@WINLINK.ORG (Note: Logistics Chief is allowed (at least initially) to use Internet as would attempt to be in functional area.) **Note -- must monitor all of the above methods of contact; may delegate. Respond to the voice tactical call sign HAM COMM**

EXAMPLE ICS-205A

PLANNING and Distributing New Information

A well-organized incident management team will be running the "Planning P" and issuing a new set of these documents perhaps every TWELVE HOURS.   Your team or your manager might need or benefit from getting this documentation -- but it isn't going to be deliverable over voice circuits and even over ham radio digital circuits, the typical documents are far too large -- so you might want to have someone with an internet connection request dumps of these documents and then reproduce the important information in TEXT format and send them to you over ham radio digital or other systems --- they can easily be sent by NBEMS, WINLINK or even non-error-corrected systems if you have a good signal.

ICS-214

This is a very important document for ham radio operators, as it is your chance to make a REPORT of what you are doing and how it is going.   You should send one in every working period!!!   You should if at all possible, be keeping it CONTEMPORANEOUSLY (as the time actually progresses!) but you may want to rewrite it for "public consumption" before you send it in at the end of the working period. They don't need to know every little detail, whereas during the day, it might have benefited YOU to have captured a load of just such little details.   Focus on making sure that your final turned-in product includes CAN

      C   Conditions --- what the situation was

      A   Actions -- what you did to further the accomplishment (or success) of your assignments

      N   Needs -- what you need in order to succeed at unfinished accomplishments

Examples:

      C   Arrived in Mexico Beach, no native radio systems operative. Assigned objective of providing LMR communications for USAR TF2 SAR operations

      A   MARC3 deployed, tower at 100'. UTAC41 on the air as Operations Net, UTAC42D in use by individual SAR Teams for tactical direct communications, VTAC12 in reserve for repeater linking, 8TAC91 on Bay County TRS linked to UTAC41 for wide area communications and voice communications to BoO [Base of operations])

      N   Additional MARC unit requested to set up in Port St Joe to provide communications support for SAR operations east of our coverage area. VTAC12 to be utilized to link this new MARC unit to our UTAC41 providing interoperable communications between TF2 and other USAR TF in the area.

(These examples come thanks to mentoring by Kevin Rulapaugh KE4NVI)

The ICS-214 is how you document to your supervisor how things are going -- and also a gentle reminder to your supervisor to SEND YOU SITUATIONAL AWARENESS DOCUMENTS!!!!

Kevin Rulapaugh tells a great story about how this worked -- in charge of a 100-foot mobile tower and repeaters to provide communications for a search and rescue group, he created an ICS-205 for those personnel based on the channels he could repeat, marked it as specific for his Unit, and sent it in to Logistics.   That got incorporated into a much larger group of ICS205's for multiple such systems in the theater, and it also helped result in him getting nightly dumps of the entire incident action plan for his area -- so he had a lot better situational awareness!   And it helped that he had high-speed digital connectivity via FIRST NET and/or other such systems.

WHEN TO USE THEM

You could certainly organize, supervise, and manage a group of volunteers using homemade documents. People do it all the time for parades, bike rides and all kinds of other festivities.   But the point of the ICS system is to have a nationwide, universally understood set of ways of managing responses -- so wouldn't it make more sense to "go with the flow" and USE these taxpayer-developed management systems and documents, so that you and your group are well attuned to them?

3   HOW TO PLAN / CREATE / CARRY OUT FULL SCALE EXERCISES AND BUILD YOUR GROUP

by Leland Gallup AA3YB

The subject of this chapter is meaningless if we don't first define what we mean by "full-scale exercise." What exactly is such an exercise, as opposed to, say, something you and your friends might want to do as a quick Saturday morning get together on the air? What differentiates a full-scale exercise from one that isn't full-scale? What does a full-scale exercise accomplish for individual amateur radio emergency communications (EMCOMM) operators and for EMCOMM groups such as local Amateur Radio Emergency Service (ARES) affiliates? And assuming the reader agrees that full-scale EMCOMM exercises might be a good idea for the local group, how do you plan, create, and carry out a full-scale exercise?

This chapter will give some tips learned from hard experience on how to run these kinds of exercises. There will also be some references offered in the course of the text and at the end of the chapter. There are many ways to create, plan, and execute full-scale exercises. What is offered here is only one way – but it's a way that works.

What you'll read here is a condensed version of the content in the Federal Emergency Management Authority's (FEMA) free online course, Incident Command System (ICS) Course IS-120 c (An Introduction to Exercises). Besides this chapter, an exercise organizer is strongly encouraged to make use of ICS 120 c – and the other FEMA ICS online courses are a goldmine of speaking the language of the national emergency response community.

Definition and Elements of a Full-Scale Exercise

Let's get some definitions and assumptions out of the way. First, we need to understand what is meant by a "full-scale exercise.' Think of all full-scale exercises as sharing some of the following elements. Considered together, these elements describe and so define what we mean by full-scale exercise. The elements include but are not limited to the following.

Goals and Objectives. Full-scale exercises are driven by goals – what you want to accomplish. Goals and objectives drive the exercise's organization and specific narrative or "story line": what happens (fictionally) in the course of the exercise.

Formality. Full-scale exercises are "formal." That is, they are planned, structured, and documented. They are not impromptu, spur of the moment get togethers on the air.

Leader. One or more amateurs/members of your group has to accept responsibility for planning, creating, documenting, overseeing, executing, and preparing the exercise after-action review. There has to be someone "in charge" or there will be no moving ahead. Besides having a leader, a formal full-scale exercise will have some kind of chain of command.

Time. These kinds of exercises will take a long time to prepare. From the inception (agreement that there will be an exercise and selection of a leader) through the actual operational period to the after-action review will probably take months. The exercise itself may run only for a short time – often only a morning. But don't think that this means the planning phase won't take months. It will.

Planning. This is more than some ideas on a napkin. Full-scale exercise planning is detailed, written, and demands attention to myriad details. The plan is written out and evolves over time. The plan usually has several component parts – not only the basic plan, but annexes such as an agreed frequency list for the participants to use during the exercise.

Personnel and Agencies. Besides the group itself, personnel for a full-scale exercise can and should involve more than just your amateur radio group. Try to get at least one – and preferably more than one – partner or served agency to "play" with your EMCOMM amateurs. These agencies are most-often governmental (such as law enforcement, fire and rescue, the local Emergency Operations Center), but for the sake of personnel decisions this can also include nongovernmental agencies, charitable organizations, or other private groups.

Logistics. A full-scale exercise involves people, equipment, and places. It is not just people and agencies that are the constituent parts of the formal exercise; also to be considered are at least two field locations from which to stage the operational narrative. Part of the logistics of site selection extends not just to looking at places, but also, permissions to use sites, equipment and facilities support, insurance, medical and other personnel support considerations.

It is easy to see from an examination of what goes in to a formal exercise that it is a lot more than just a spur of the moment bit of radio fun. That being said, why should you and your group bother? Simply put, because doing a formal exercise develops individual and group skill in ways that are absolutely critical for the amateur's relevance to 21st century emergency radio communications – one of the two principal reasons why amateur radio exists. Besides, formal exercises are actually fun and rewarding. We are amateurs, after all, and love this for the great hobby that it is. If you accept this, let's turn now to how you create the exercise.

Creating and Executing the Exercise

Goals. Simply put: you need to decide what it is that you want your group to practice and get better at by conducting formal exercises. The North Florida Amateur Radio Club (NFARC) is an ARRL affiliated club in the Gainesville, FL, area. Its purpose is wholly to support the ARES program in the Alachua and surrounding counties areas of northern central Florida. NFARC members decide on an annual basis what should be the clubs' exercise goals for the upcoming year. NFARC's exercises are created specifically with these goals in mind. This is an example of club goals for 2018 in the form of capabilities that NFARC members want to work on:

Consider the following as a listing of some goals your group may want to adopt and then put in to practice by way of exercise creation.

a)   Mobile Communication assets/skills -- VHF and HF

b)   Skills at using vacuum tube gear for EMP, to make longer-range comms out of county

c)   EOC comms out to the community

d)   Becoming better known -- making our capabilities more known to possible clients

e)   Microwave capabilities

f)   Short message communications

g)   FEMA forms abilities to transfer (mentioned were FLMSG)

h)   Last mile communications

i)   Traffic sending ability -- discussed were voice, DTN, PSK31/ FLMSG

Technical/Engineering Skills.

Table 1-1 includes some technical, equipment-related, goals worthy of consideration. There are as many possible goals as your group can think of that relate to what you might want to practice.

Table 1-1

Antenna building	This includes the rapid building of field-expedient antennas from materials likely to be found on scene in the field, using simple wire, cord, and insulators. Antennas can be VHF/UHF or HF.
Soldering proficiency	Important for repairing or creating equipment in the field.
Equipment familiarization	Expose group members to gear used by other members or at locations likely to be used that may have radio equipment. For example, there is an amateur radio room in the Alachua County Emergency Operations Center with commercial HF equipment that can be operated by amateurs on the ham bands. This is not typical "ham" gear, and hams must be able to operate the gear they are likely to encounter in emergencies.
Equipment deployment	Rapid set up and take down of antennas, transceivers, computers, power systems, and interconnecting cables. Essential skillset that must be practiced. This is harder than it seems, since field conditions can require unsupported antennas (no trees to hang wires from), long cable runs, and awkward operating positions.

.

Operating and Procedural Skills.

Table 1-2 includes various operational aspects of EMCOMM that your group can exercise. These are just some of many...your local group and conditions dictate the possibilities.

Table 1-2

Operating frequencies	Variety of HF frequencies; VHF (mostly for simplex and repeater FM and digital/packet; UHF (simplex and via repeaters); microwave. All these frequencies can be selected as goals for exercise use. The best way to become familiar with and more agile operators on all the frequencies and modes your groups' privileges allow is to use them. If most of your members are Techs, focus on VHF and UHF for your exercise goals. In any event, look at frequencies your members can use and use them – as many of them as you can.
Operating modes	**Voice:** SSB, FM, using simplex and repeaters as appropriate. Getting over "mike fright" in new hams is something that an exercise can overcome. **CW:** (if any of your members can do this). **RTTY:** an underused mode that is ripe for exploring through an exercise. **Digital:** These modes are becoming extremely important in emergency communications. With a group or members unused to digital modes, start simple. Start with VHF packet. They can use Winlink for this. This will probably require members to build or purchase a soundcard interface (such as the Signalink USB). The interface and a computer are necessary, but that's what is necessary beyond most transceivers for digital operations. PSK-31 on HF is a good and simple digital mode on these bands so members can become familiar. The next step is to have operators develop skill with Winlink's various HF modes, such as Winmor, Ardop, Vara, and Pactor (if your members have access to a Pactor modem). Just sending and receiving a digital message may be one goal (besides phone) for a first full-scale exercise.
Nets	Many amateurs are already familiar with nets, but actually using a net to do something productive other than merely to check-in is a critical EMCOMM capability. Moreover there are differing nets for differing modes. An excellent exercise objective would be to establish a net through use of predetermined frequencies and then use relays to pass information to a net control. More challenging would be to use the SARNET (Florida-specific statewide DOT sponsored amateur UHF net) as a net extender.
Traffic	Handling traffic is easy in principle and difficult in practice. Simple FM phone traffic over handhelds, with the requirement that exercise participants send and receive a Radiogram, is a superb, inexpensive, rewarding, and inexpensive exercise goal!
Incident Command	No emergency communications exercise in 2019 should fail to use FEMA's

System forms	Incident Command System (ICS) and its documentation structure. ICS familiarity – certainly with often used forms – is a key part of interoperability between your group and the wider world of the agencies you want to serve. The ICS system uses a variety of forms. These forms can serve as the foundation for how you create, plan, execute, and document what happens in your exercise. Documentation is essential to a good full-scale exercise. ICS forms will be discussed at greater length below, but there are some specific forms can be used by all participants in an exercise. Using these forms is a process skill; so a great exercise goal is to have all participants become familiar with and use **ICS Form 201** (Incident Briefing), **ICS Form 205** (Incident Communications Plan), and **ICS Form 214** (Activity Log). Examples of the 201 and 205 will be provided below. You can find Word-fillable versions of these forms can be found on Google; one such source is at https://www.fda.gov/EmergencyPreparedness/NIMS/ucm268797.htm The Food and Drug Administration uses the ones at the link, but they are easily modified for your purposes. Using Word forms, and agreeing on specific word-processing and spreadsheet formats for your exercise creation will allow commonality. That means your group can use just one program type (for example, Open Document) to create exercise documents. Agree on standard document types and you will save yourself trouble.

.

Creating and Executing the Exercise. After your group's capability objectives are established, then select your goals for an exercise. These are the skills, knowledge, and abilities that you want to further by way of an exercise. So where do you start? Where do you begin conceiving of an exercise? After this section we will look at a suggested timeline for how to organize events up to and after the exercise.

The ICS 201. Very useful not only for drafting the exercise scenario but also for coming to grips with other details of your exercise is the **ICS Form 201**, Incident Command Brief. Draft an ICS at the very beginning of your creative process. It provides a structure that in itself a useful template for exercise creation. Besides, an early version of an ICS 201 with the incident narrative and locations sketched out is very useful for "propaganda.". Since it is increasingly likely your served agencies "speak" ICS; using an ICS demonstrates your commitment to interoperability by employing documentation your target partner agency may already be very familiar with. Speak a language they understand and they're much more likely to work with you, not only during an exercise but also during an emergency. Don't worry about using early version of the ICS for this purpose.

Figure 1-1 is an ICS 201 used in late spring 2018 by NFARC. This 201 shows the simple and short scenario for the exercise: the "who" that will participate; the "where" of exercise locations; and the "what and when" is the timeline of what will unfold during the exercise.

Incident Briefing (ICS 201)

1. Incident Name: **Simulated Emergency Test 2018**	2. Incident **Number:** **4**	3. **Date/Time Initiated:** Date: 10/13/2018 Time: 0900local

4. Map/Sketch (include sketch, showing the total area of operations, the incident site/area, impacted and threatened areas, overflight results, trajectories, impacted shorelines, or other graphics depicting situational status and resource assignment):

Santa Fe College/Perry Construction Institute. IC and Comms 1 Location

NW 91st St, Gainesville, FL; 1st left southbound (from 39th Ave NW), Perry Inst on right; IC and Comms 1 on left in open field with tree line on east side.

[Copyrighted satellite photo deleted: Illustrated location of Comm 1, in a field at the Santa Fe College Campus, relevant streets, North / South]

Gainesville Senior Recreation Center. Comms 2 Location.

5701 NW 34th Blvd, Gainesville, FL 32653 (352) 265-9040; across from Walmart on 121; northbound right turn in to parking lot of Senior Center.

[Copyrighted satellite photo removed, that illustrated the location of Comm and relevant roads, and parking.]

Alachua County Emergency Operations Center. Comms 3 Location.
1100 SE 27yh St, Gainesville, FL (352) 264-6500. SE Hawthorne Rd; on right hand side while southbound. EOC is in the back of the parking lot beyond the Sheriff's Department offices.

[Copyrighted satellite photo removed: Showed location of Alachua County Emergency Operations Center, relevant roads and parking.]

5. Situation Summary and Health and Safety Briefing (for briefings or transfer of command): Recognize potential incident Health and Safety Hazards and develop necessary measures (remove hazard, provide personal protective equipment, warn people of the hazard) to protect responders from those hazards.

Situation Summary:

There is a wide-spread infrastructure failure. Grid electrical power, the internet, and the satellites do not work. It does not seem to be an EMP, because batteries, solar, and generators work, and can be used for field and EOC comms.

For the purpose of this exercise scenario, a local zoo has had an escape. The zoo's electromagnetic cage locks released when the power went down. An unknown number of wild animals, some potentially dangerous,

1. Incident Name: **Simulated Emergency Test 2018**	2. Incident **Number:** 4	3. Date/Time Initiated: Date: 10/13/2018 Time: 0900local

have gotten loose. A team of search and capture officers is on the scene near Santa Fe. They are not HAMs. Moreover, the local MARC unit has deployed to the site to provide Alachua County local government communications infrastructure.

A search & capture team has deployed and taken some local residents (student actors and others) to a shelter to stay out of harm's way. Among these are persons with medical conditions requiring electricity for care.

Exercise note: Do not forget to announce that your transmissions are part of an exercise.

Health and safety:

- Nearest local hospital to Incident Command is UF Emergency Shands at 8475 NW 39th Ave.
- Nearest local medical care to the Senior Center is CareSpot at 3925 NW 43rd St.
- Nearest local medical care to the EOC is UF Shands at 1515 SW Archer Rd.
- The weather is likely to be hot and humid. If outdoors, wear sunscreen and drink plenty of water.
- Gainesville has insects that can carry human pathogens. If in areas that are prone to mosquitoes and ticks, wear repellent.

6. Prepared by: Name: S. Halbert/L. Gallup____ Position/Title: NFARC/ARES_____ Signature: _____

ICS 201, Page 1	Date/Time: 09/04/2018

Figure 1-1   Example ICS-201, Initial Page

Scenario. Full-scale exercises require a causative event – something like a disaster that causes a need for amateur radio assistance. Another cause could be a breakdown in normal communication modes. From the starting point of such a fictional need or disaster the exercise creator drafts a scenario with "injects" (more on this below). The scenario is the story line of how the disaster plays out in the exercise period, and the steps that are taken in the exercise to advance both the narrative of the disaster and the goals that are to be achieved.

The beginning of any exercise must take into account a key bit of conventional wisdom: all disasters are local. The prevailing weather, geography, geology, terrain, fire/flood likelihood, population, and history of your location should drive the creation of a scenario that will be one that you'll likely encounter. Weather and fire are not the only disaster causes that you can think about. In a hyper-interconnected world the disaster causes could be power outages caused by electromagnetic pulses or cyber activity.

Here in Florida earthquakes are not really a concern. Nor are tsunamis. But hurricanes and tornadoes certainly are. If you are "dreaming up" an exercise scenario in Florida you should probably begin with a hurricane or tornado as the cause of a disaster. In Alachua County we ran exercises using hurricanes and wildfires as scenario-themes. The differing themes led to different problems to solve; nevertheless, many times the exercise goals were similar.

Who – Persons and Organizations for the Exercise. Exercise participants should include both amateur radio operators, to be sure – but wait, there's more: representatives from at least one public service agency or non-governmental/private organization. It's not a full-scale exercise if there isn't outreach to non-amateur element: community law enforcement, fire and rescue services, or for that matter cooperative state services (such as state park officers and forest service personnel). You can reach out and recruit county and municipal emergency services – such as EOCs or those responsible for setting up emergency shelters in the event of inclement weather. Anyone you can think of who you could think of that might actually benefit from amateur emergency communications.

Amateurs operators. Recruit amateurs by getting up in meetings of your ARES or local radio club and telling them that you would like to reinforce the club's strengths and have a lot of fun by doing an exercise. Use an example of one that's been done and that was fairly simple. The Wacassassa Wildfire exercise (see the ICS 201 above) is a perfect example of something you could use as a talking aid for a club presentation. Then follow up with emails/calls to members. Try to gather everyone you can – from Techs with handhelds to Extras with towers. Everybody. And as many differing skills levels with a variety of operating modes as you can. This is an opportunity for your Techs to actually get involved and do something fun – it's also great exposure to the uses of General and Extra level privileges.

Think about getting non-amateurs – family members, Scouts, you name it, to participate as actors in your disaster/emergency scenario. This is great ham outreach; but an even better outreach opportunity is to engage them as "players" by use of FRS walkie talkies to contact FRS-equipped hams who can then inject traffic or information through the amateur net you'll establish. The FRS non-amateur can then see, through your exercise, how amateur radio works and may be a potential ham recruit!.

Served agencies and organizations. In your locale, these are the local governmental agencies – municipal, county, and state, that would most likely be able to use amateur operators' help in times of disaster. Examples are these have already been touched on, but it bears repeating that these agencies are local police, fire and rescue services, emergency service providers (such as designated shelters), and emergency operations centers.

Evaluators. Try to get some evaluators – people whom you give the entire exercise package, and who watch and listen to what happens. Their job is to give a critique of what went right and what went wrong from the "outsider" point of view.

What and Where. Use your creative powers to think about what you would want your EMCOMM hams and the served agencies to do. And this leads to the "where" that you'd like them to do it. The ICS 201 sample above shows that this can be quite simple. In the Wacassassa exercise, the scenario envisaged a need for hams to cooperate with governmental agencies: in this case, the Alachua County Fire and Rescue and the Florida State Forest Service. Take a look at paragraph 5 of the ICS 201 for the Wacassassa exercise, hopefully as inspiration.

For locations, it was again quite simple. There were three locations from which hams and the served agencies operated radios. Two were only 200 yards apart. These were the "field" locations. Hams and served agency personnel set up field VHF and HF radio stations as they would be set up for real – however antennas could be erected at the locations, and with whatever emergency power sources (batteries and generators) were at hand. Frequencies were those for both amateurs and served agencies, consistent with their frequency permissions.

The third location was the county Emergency Operations Center. If there is any one emergency service facility in your area with which hams should have a working relationship, it is the EOC. If you haven't already, high on your group's list of enduring goals should be establishing a relationship with your local EOC – better still if you can get antenna and radios there that amateurs can use. In Alachua County the EOC has a radio room specifically for amateurs, and both VHF and HF antennas for amateurs. Develop and nurture the human contacts with EOC personnel. If you do this now, then when the need is real the EOC will know what hams can do, and the amateurs can fall right in with the overall disaster assistance scheme.

When. This is the operational period of the exercise. Look at paragraph 8 of the Wacassassa 201, the "Current and Planned Actions." Your 201's paragraph 8 is the meat of the exercise. It will detail how, where, and when personnel check in, the orientation briefing, inventory of equipment, safety briefing, deployment to operational sites (if necessary), and establishment of radio communications among your participants, be they afoot with HT's, mobile, at "fixed" field locations, or at a hardstand site such as an EOC or building designated as a shelter.

How – the Frequencies. Your must have an agreed list of frequencies for your exercise to work. The chosen frequencies can include not just amateur band frequencies, but also other radio frequencies, such s FRS, if your scenario calls for them. Here again, the ICS system of forms is the way to go, and in this case it is the **ICS-205**. **Figure 1-2** shows the communication plan for the Alachua County ARES group, and it is a great sample to use for an exercise 205.

L i n e	Ch #	Z o n e	Function	Channel Name/ TG Name	Assignment	RX FREQ N / W / SSB	RX Tone/ NAC	TX Freq N or W	Tx Tone/ NAC	Mode A, D or M	Remarks
			INCIDENT RADIO COMMUNICATIONS PLAN ICS-205		Incident Name Alachua County Emergency		Date/Time Prepared 07/18/18			Operational Period Date/Time		
1			Tactical	K4GNV82Repeater	Amateur	146.8200 W	123	146.2200 W	123	A	Primary / Command net
2			Tactical	K4GNV685Repeater	Amateur	146.6850 W	123	146.085 W	123	A	Secondary Repeater
3			Tactical	2M 146.490 SIMPLEX	Amateur	146.4900 W	None	146.4900 W	None	A	Simplex Local Comms-1
4				GNV-PACKET070	Amateur	145.0700 W	CSQ	145.0700 W	CSQ	D	VHF Digital – EasyTerm / WINLINK
5			EMAIL	KX4Z-WINLINK	Amateur	USB DIAL Frequencies 3594.0 7102.0 10140.0 14096.5	None	USB CENTER FREQ 3595.5 7103.5 10141.5 14098.0	None	D	HF WINLINK – Local
6			EMAIL	WINLINK HF	Amateur	TBD	None	TBD	None	D	HF WINLINK – National
7			HF VOICE	80M LOCAL	Amateur	3950 LSB	None	3950 LSB	None	A	LOCAL if NFAN down
8			HF VOICE	NFAN	Amateur	3.950 7.242 7.247 LSB	None	3.950 7.242 7.247 LSB	None	A	North Florida ARES
9			NEIGHBOR	FRS2GMRS	Anyone	462.58750	None	462.58750	None	A	HAM NEIGHBOR WATCH
10				OCALA-GNV PKT	Amateur	145.0300 W	CSQ	145.0300 W	None	D	Digi to KX4Z-10 via W4DFU-8
11			Tactical	W4DFU-91Repeater	Amateur	146.910 W	123	146.310 W	None	A	Tertiary Repeater
12			Tactical	NAT 2M CALLING	Amateur	146.5200 W	None	146.5200 W	None	A	2 meter calling frequency
13				Select Channel		Mobile Rx	Rx Tone	Mobile Tx	Tx Tone	A / D	
14				Select Channel		Mobile Rx	Rx Tone	Mobile Tx	Tx Tone	A / D	
15				Select Channel		Mobile Rx	Rx Tone	Mobile Tx	Tx Tone	A / D	
16				Select Channel		Mobile Rx	Rx Tone	Mobile Tx	Tx Tone	A / D	
17				Select Channel		Mobile Rx	Rx Tone	Mobile Tx	Tx Tone	A / D	
18				Select Channel		Mobile Rx	Rx Tone	Mobile Tx	Tx Tone	A / D	
19				Select Channel		Mobile Rx	Rx Tone	Mobile Tx	Tx Tone	A / D	
20				Select Channel		Mobile Rx	Rx Tone	Mobile Tx	Tx Tone	A / D	

Prepared By (Communications Unit)

Incident Location

County ALACHUA   State FL   Latitude   N   Longitude   W

The convention calls for frequency lists to show five digits after the decimal place, followed by either an "N" or a "W", depending on whether the frequency is narrow or wide band.

Alachua County ICS-205 for 2018   [Screen capture, G. Gibby.]

Envelopes. If you read the Wacassassa paragraph 8, you saw "envelopes" mentioned. What are these? They are "secrets" that the exercise organizer drafts and **injects** in the play at specific points in the exercise. The envelopes are drafted in advance and literally placed in envelopes to be opened by specific participants at specific times. These are also known as **MSEL's**, from "Master Event Scenario List." The MSELs are what gets things going – they will have the information or messages that are to be sent, the problems that participants must deal with (such as the repeaters are now out), and similar surprises. MSEL injects promote adaptability and flexibility, but they are also the essential information that must be passed along or dealt with in order for the exercise to work and advance your group's goals.

After the Exercise. The job is not done until the paperwork is complete.

Hot wash. This is held right after the exercise, at a non-threatening location, such as a restaurant (with reservations in advance) for the entire body of participants. Ask everyone to list what they think went right and wrong; then ask the evaluators for the same thing. If you don't do this immediately after the exercise much of the immediate details will be lost to the fog of forgetfulness. The organizer or a designated scribe should take notes of what was said.

After Action Report. Even if a hot wash is done, and notes are kept of the bottom line – whether and how the exercise contributed to the group's objectives and goals, a more formal After Action Report (AAR) in written form is very useful. The AAR can take the form of the various publications of the NFARC group as produced by Dr. Gordon Gibby, MD, KX4Z, or it can be less expansive. AARs are great not only because they really get in to the heart of a critical analysis of success and failure, but because too they are a superb resource for future exercise planning. No reason to reinvent the wheel if you can remember how the wheel is made. Be thorough, be honest, and be respectful in your AAR. But do an AAR in writing. The AAR can contain all the ICS forms used in the exercise, the MSEL injects, the candid notes from the Hot Wash, and photographs of participants during play (highly recommended!).

Timeline: Sequence of Planning Events. The exercise creator and organizer can can should delegate as much of the following as possible – otherwise the effort can seem overwhelming. The day counters are just markers. Progress is obviously made continuously all the way up to the day of the exercise.

120 Days Before Exercise (BE). .

Date for Exercise. Look at your community and look at your calendar. It is not a bad idea to run an exercise with a hurricane scenario in the hurricane season (August-October), but be aware that reality can hijack your plans and make personnel unavailable. Select a Saturday that isn't part of a long holiday weekend when people typically go away. When you've got a possible date, run it against a calendar of public events. For example, don't try to run an exercise in Alachua County on a Saturday with a Florida Gators home game! Deconflict your dates with as many such events as you can.

ICS 201. Draft your first exercise scenario and your first 201.

Agencies. Use your first ICS as an entry point in to the served agency(ies) that you would like to have work with your group in the exercise. If you need to sell your group and what they can bring to the table by way of assistance in disaster preparedness, begin earlier than 120 days – but for sure after this point look at recruiting one or more agency or non-governmental organization. Getting the served agency or EOC lined up early is even more important that getting your own people, because the agency might have conflicting training requirements or events that could interfere with their ability to participate. They are the professionals. We are the amateurs. We serve them.

Locations. Begin scouting out locations that make sense in view of your idea for the scenario's narrative.

Personnel. At least think about who among your members might be recruited to do what. What are their strengths and weaknesses? Who among them are reliable and good leaders? Who among them have the class privileges you're seeking to exploit (for example, HF work). Who can do CW? Which of them are can bring their own field stations? Which can do digital modes? It is important that you get a feeling for the people you can approach before you approach them. And when you've done this initial organization/task list assignment that matches your concept of the 201, then make your first attempt at recruitment.

90 Days BE.

Locations. For those locations already agreeing, send out confirmation emails or phone calls. Of course, if you've not found a good place, keep looking!

Permissions. Along with the actual locations, you will need to make sure that whatever the owner of the location requires by way of permission documentation is started. For example, the Santa Fe College here in Alachua County, site of many or our exercises, including this Symposium, requires a written request for facilities use. Facilities/location use isn't necessarily as simple as an oral agreement.

Insurance. Another fly in the permission ointment is that the location owner may require proof of insurance indemnifying them against liability and damage. In the case of the Alachua ARES group, we founded the North Florida Amateur Radio Club and affiliated with ARRL so that we could obtain liability insurance through the latter. Remember, ARES is a program, not a club; your group should be with a club or something that an insurer will recognize to get insurance. Insurance is not expensive, but is always a good idea for your group's efforts. If none of your location owners require written permission or insurance, fine. But just be prepared to follow this through if the owner does want insurance coverage.

Personnel. Keep up your search for people to agree to take particular roles in your exercise; as they do, revise the first 201.

ICS 201. Edit the 201 to show an organizational scheme – who does what and where – as you secure participants and locations.

ICS 205. By now you can draft a first cut on your exercise 205.

Publicity. Determine what needs to be done to publicize your exercise in your ARRL section's websites. Same, of course, with respect to your local club website.

60 Days BE.

Rehearsal of Concept (ROC Drill). Without going through a dress rehearsal, at least have a good idea whether the concept will work at all. Discuss with your key participants to see if your or they see any show stoppers or things that need to be corrected.

Personnel and Locations. With two months to go you should be closing in on a good idea of who will be available and what locations might be used. Location permissions if necessary should be in progress, as should the securing of insurance for your group – again if needed.

ICS 201 and 205. Edit and change to flesh out personnel and locations, and to fine-tune the timeline for the exercise period.

Envelopes. Begin drafting your MSEL injects and bounce them off a non-participant to see if they make sense. Share with your evaluators.

Rehearsal planning. Begin planning and publicizing a table-top rehearsal for 30 days BE. Unlike a ROC drill, a table top should involve actual set up of stations, run through of equipment and familiarization, and if possible quick set up to test for real the concept. If possible, do this, too, at at hard stand locations such as the EOC and places such as schools and community centers that are used as designated shelters. Just don't make it as big a production as the actual exercise.

30 Days BE.

Rehearsal. This is the time to get your strongest players locked down as to what equipment they have and can bring to the actual location; that they can set up and operate in all modes called for in the exercise scenario; and that they can use the ICS 205 frequency list that you have drafted for the exercise.

Personnel. Following the rehearsal you should have a pretty good handle on your key participants and their positions in the exercise. Reach out again to your served agencies and get affirmatives on their participation. Be flexible, as participants may drop out and have to be replaced.

ICS 201 and 205. In light of the dress rehearsal/tabletop results, revise the ICS 201 and 205.

Publicity. Announce on your local ARES nets and any pertinent website that the full scale exercise will take place in a month. Periodic reminders as you count down the remaining weeks are a good idea.

15 Days BE.

Media Publicity. Contact local media such as TV and printed press and submit a press release about your exercise. Think of what you likely do for Field Day publicity; the advantage here is that there is a public-service aspect to this that is different from Field Day, and that local non-amateur organizations are involved.

Reminders. Again, through nets or emails and phone calls.

7 Days BE and Closing.

Personnel. Begin adjusting for people dropping out or being added. Now is the time when your flexibility and patience will be most tested.

ICS 201 and 205. Final versions. Place on your club's website or other publicly accessible portal.

Remaining ICS forms. Finalize and publish any other ICS forms you may care to generate, such as **ICS 206** (Medical), and **ICS 211** (Check-Ins). Be prepared to use the latter to check all participants in for the actual exercise.

Locations. Final coordination with site owners to make sure your planned sites will indeed be available!

Publicity. Within a couple of days of the event, make sure that your press release or coordination with (for example) local newspapers and TV are reiterated. Frequent repetition makes for frequent success.

BE. The big day. Start with an orientation briefing at a single location, and do the incident check-in and safety briefing. Make sure you have hard copies of your ICS 205 Communications Plan for all participants. Encourage use of the ICS 214 Activity Log if you have receptive players. Finally, there is no reason you yourself cannot participate as a player. Evaluators should just evaluate; they should not play. Encourage everyone to take pictures with their cell phones, as these are a wonderful aid to memory and make the after action reports much easier to understand. Pictures are worth more than the proverbial thousand words.

Release the hounds and have fun!

Hotwash. Very important to reinforcing what went right and what didn't. Have someone act as a scribe to record the thoughts of participants before they scatter.

Formal AAR/Publication. With the photos taken by group members, and their permission to release, use for postings to, for example, the ARRL Section web page and your own club – a short article, if you are not inclined to do a publication, is a good version of a "quickie" AAR.

The formal, full-scale, exercise is fun, builds your group, and sets you up for success in what we're trying to do – provide EMCOMM for our communities when they most need us. Now go out there and do it.

4 LESSONS FROM CASCADIA RISING AND OREGON 2018 SET

by Gordon Gibby KX4Z

CASCADIA RISING

Good example of recent FEMA / Amateur Radio Coordination in 2016 Full Scale Exercise
- Multi-state exercise planned over 2 years
- Involved over 20,000 people
- Lasted 4 days
- Modeled the aftermath of an 800 mile long catastrophic earthquake in the Cascadia Subduction Zone
- World wide activation expected to be necessary for rescue effort
- Pulitzer price winning 2015 essay about this eventual catastrophe

FEMA After-action Report almost gushes about how well the amateur radio component worked

> *Observation 1.2: Strength:* ***Amateur radio was a critical mechanism for backup communications.*** *Analysis: Numerous jurisdictions utilized amateur radio effectively to coordinate in a communications-degraded environment. For some jurisdictions, this exercise marked the first time public messaging was issued via amateur radio. This exercise served as an excellent opportunity to train novice amateur radio operators and provided experience to all operators in federal, state, and local amateur radio integration....*
>> https://www.fema.gov/media-library-data/1484078710188-2e6b753f3f9c6037dd22922cde32e3dd/CR16_AAR_508.pdf

Extensive use of both HF (primarily) and VHF Amateur Radio Capabilities
- 80 meter / 60 meter operations lauded
- EXTENSIVE use of WINLINK systems – **so much so that over-reliance over them was concerning**

- Idaho ARES members with HF go-box & Chameleon antenna were considered more effective than Idaho national guard comms
- WINLINK messages flowing into Oregon centers, and to ARRL Headquarters in Conn.
- WINLINK **gateways in Nevada and California were leveraged** (the HF Advantage)
- Majority of traffic over WINLINK (would primarily be HF)
- Tsunami warnings, earthquake warnings, etc. were the first messages
- 20 meter SSB nets also utilized (HF)
- **Pointed out how much more efficient WINLINK digital was compared to voice.**

- Biggest concerns: **not enough amateurs are trained and available to provide effective response. Estimates in one state: 2-3% of licensed amateurs.**
- ARRL after-action report almost reads like advertising copy for WINLINK system....

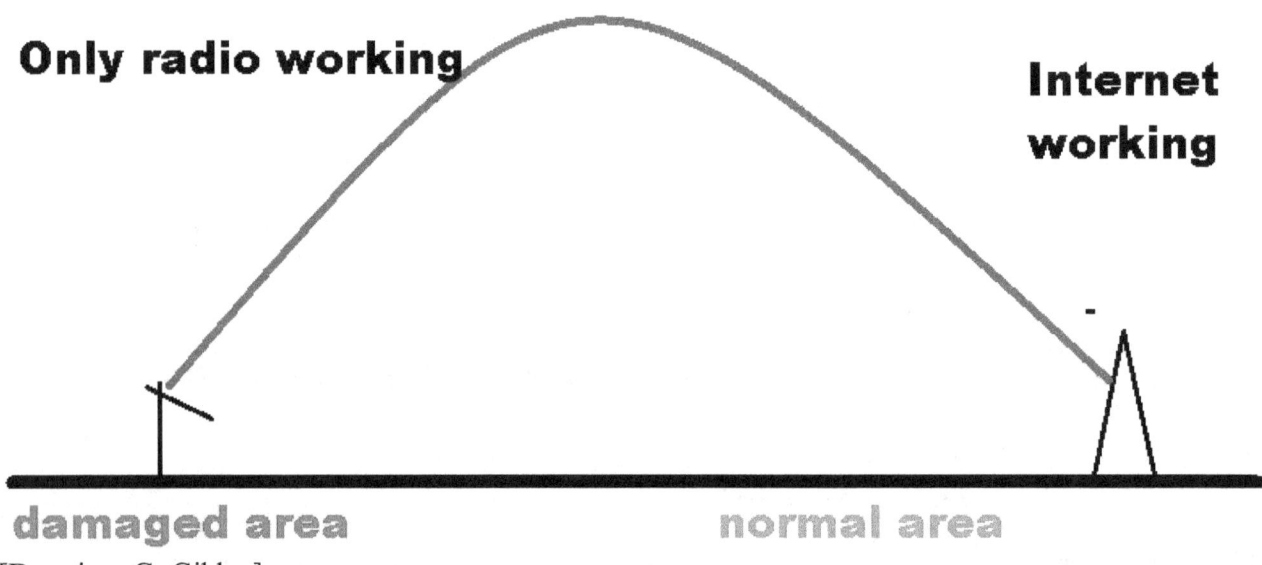

[Drawing, G. Gibby.]

NOTE: THE CASCADIA RISING FEMA TEST UTILIZED THE OPTIMAL SETUP FOR WINLINK -- COMMUNICATE DIRECTLY OUT OF THE DAMAGED AREA TO UNDAMAGED STATIONS WHO CAN FORWARD COMMUNICATIONS EASILY (AUTOMATIC INTERNET TRANSMISSION) This mimics its original design, to serve ships at sea, connecting to internet-connected land stations.

To see how modern amateur radio is fitting into major FEMA / DHS planning, you can read the ARRL and FEMA After Action Reports on the Cascadia Rising 2016 Full Scale Exercise --- which covered multiple STATES not just one city. This can better explain the roles that both shortwave (HF) and VHF digital play in a truly large disaster response, and these published reports give up to date information on the critical skills and issues involved.

Cascadia Rising was an attempt to model and simulate the outcome of an 800 mile long full rip earthquake in the extremely dangerous Cascadia Subduction Zone. Such a catastrophic event has been written about extensively, gaining notoriety after a Pulitzer prize winning 2015 essay about its astonishing destructive potential. *It is expected that a WORLD WIDE response would be needed in the recovery effort.* Tsunami generation would be enormous. I get the general idea that hundreds of thousands to millions might die. This was the first huge scale effort to prepare for this event.

TWENTY THOUSAND people participated in this full scale exercise carried out June 7-10, 2016.

The FEMA after action report is here:
https://www.fema.gov/media-library-data/1484078710188-2e6b753f3f9c6037dd22922cde32e3dd/CR16_AAR_508.pdf

 Ham radio -- and WINLINK -- got a glowing report.

The ARRL after action report is much juicier reading if you want to really see what worked, didn't and needed improvement on our end (communications). You can read that here:
http://www.arrl.org/files/file/Public%20Service/ARES/Cascadia%20Rising%202016%20-%20Final%20Report.pdf

I'll call attention to some really important statements in each report. I cannot hope to give you a full flavored and balanced view of all the amateur participation in just a few paragraphs. Both HF and VHF communications were vital, in different ways. What comes through loud and clear however, in these reports is that HIGH FREQUENCY COMMUNICATIONS WERE CRUCIAL. They are mentioned FAR FAR more often than anything else. When you are dealing with a REAL catastrophe, your problem isn't to communicate 10 miles --- **it is to reach people hundreds or thousands of miles away** ***who still have intact capabilities to help.***

FEMA: *Observation 1.2: Strength: Amateur radio was a critical mechanism for backup communications. Analysis: Numerous jurisdictions utilized amateur radio effectively to coordinate in a communications-degraded environment. For some jurisdictions, this exercise marked the first time public messaging was issued via amateur radio. This exercise served as an excellent opportunity to train novice amateur radio operators and provided experience to all operators in federal, state, and local amateur radio integration....*

ARRL: [Note: the "Chameleon" is an end-fed HF antenna]

> *For this specific scenario, it turns out a radio amateur with a "Go Box" was more useful than the Joint Incident Site Communications Capability (JISCC). We are more mobile, have better frequency coverage, and have better situational awareness. That said, the JISCC has capabilities to communicate with other government agencies that Amateur Radio doesn't have. This deployment caught the Idaho National Guard (IDNG) between major personnel changes, and it was great training for them. The exercise also gave them a basis on which to train and grow. They had issues with a loss of satellite communications, were unable to communicate with the aircraft, and had very poor HF reception due to a poor antenna. Our ARES member with a Go Box using a Chameleon sloper had much greater capability. The IDNG was impressed.*

Note: the ARRL report mentions WINLINK so many times you would have thought it was ADVERTISING COPY for winlink.....and in general, they are talking HF, but there is also VHF winlink, just as we have developed locally.

ARRL:

> *Protocols were not always followed (e.g. use of 146.540 simplex for emergency voice traffic). There were too many voice requests to send traffic that could have been sent by Winlink.*

ARRL:

> ***Puget Sound Energy's largest concern is the heavy reliance on Winlink.*** *If Winlink is available, it obviously should be utilized as much as possible. However, we believe the availability will be severely limited in an event of this magnitude. Our concerns are: o **Most if not all Radio Mail Server (RMS) stations are not located in hardened sites.** Many, if not all, would be out of service due to damage or lack of power, especially for prolonged time, as would be in a disaster like this. We realize that, for the sake of the exercise, the assumptions made concerning power supply were not realistic. It should be stressed that conclusions made in this regard should be viewed realistically, in order not to skew expectations in a real event, and not to get a false sense of security.*

ARRL:

> *In an effort to take advantage of preformatted messages, Winlink should be the standard messaging system. However, some jurisdictions have been using Fldigi and others across the state may not have data-handling capabilities. In order to lessen the potential for confusion, Winlink should be adopted as the standard for sharing messages intra- and interstate.*

ARRL:

> *At the National level, Oregon ARRL Liaison Station KX7YT began transmitting Winlink traffic by HF Winmor to ARRL Headquarters station W1AW in Connecticut, using gateways in Nevada and California. The first messages were the exercise NOAA National Tsunami Warning Station earthquake notification and tsunami warnings. Over the course of the morning, 17 messages were sent/received by W1AW. ARRL HQ was contacted on 20-meter SSB and check-ins with the ARRL 20-meter Cascadia Net were made.*

ARRL:

> *About 100 Winlink messages were received at Oregon OEM from participating agencies that chose to be active on Wednesday. Again, at 0800 local time on Thursday morning, unit activity Winlink traffic started coming into Oregon OEM, and County units began checking into the 80-meter net. Oregon ARES operators took traffic on 60 meters, SHARES and FNARS, but the majority of the activity was again on HF Winlink*

ARRL:

> ***Oregon ARES/RACES has far too few active, trained operators. This is a system-wide problem.*** *Of the 17,500 licensed Amateurs in Oregon, only 2 – 3% are actively involved in ARES. Should a real disaster occur, ARES would be stretched beyond its capacity. This is as much a problem for all of the Emergency Managers we serve as it is for Oregon ARES/RACES. Joint action from all stakeholders is needed to address this issue. Once recruited, actually training the numbers of people needed is another problem to address. • More HF NCS stations must be recruited and on the air during exercises. As noise levels are very high at OOEM, the ARU*

operators (especially on 80 meters) have a very difficult time hearing others. • Equipment: While HF Pactor 3 works well, we would benefit from Pactor 4 capabilities to deal with the volume of traffic coming into our County and State EOCs. Further, the existing Oregon ARES Digital Network was funded in 2008 and is now 8 years old. At some point, funding to begin replacing the existing equipment will be needed.

In another place in the ARRL report a writer said what was optimal seemed to be 2 voice operators and 2 digital operators at each emplacement. In another report, there was ONE heavy airplane-capable airport expected to be remaining in the damaged states --- and WINLINK HF turned out to be the way to get comms in and out of that area and one ham was the key to success at that airfield.

If you're going to be maximally effective you can't just be a "voice guy" or a "digital guy" or a "VHF guy" or a "HF guy"......if you want to be maximally effective you need to be a HAM who knows how to do ALL KINDS OF THINGS. Then you can pick and choose how to get the job done when things around you are falling apart. A one-trick pony will not cut it many times. And when you are dealing with really BIG problems, I'm trying so very very hard to convince you: you absolutely MUST have high frequency communications and you really better have digital ones also.

The OREGON 2018 SET provides an important contrast to the findings of the FEMA Exercise....

OREGON 2018 SET

http://www.oregonaresraces.org/wp-content/uploads/2018/11/ARES-SET-Fall-2018-After-Action-Report-04.pdf

[All quotes unless otherwise notated are from that document.]
This was a much smaller exercise (one state, just hams) designed by a very active club, which had the following goals

"a. Test county station's ability to use Winlink by radio under moderate conditions
b. Test OEM's ability to receive and reply to approximately 20 "text only" Winlink messages by "radio only" in a one hour period of time
c. Set up and test a network of control and relay stations for SSB on 75M
d. Test counties abilities to connect to the network of stations on SSB 75M
e. Test the creativity of aux comm units to get a message out in a situation that is out of the ordinary "

"There are five tasks in the SET.
1. participating counties send one Winlink message (Oregon, Situation Report SITREP) to OEM without the use of the internet. (relays allowed, may go via HF or VHF) (emphasis added)
2. OEM will reply to each Winlink message without the benefit of the internet (neither using telnet nor the internet connected to the OEM gateway)(relays are allowed).
3. participating counties will receive the OEM reply without the internet
4. participating counties (and individuals) will check into the OEM net on 3.964 SSB
5. counties may optionally participate in the MacGyver task (no OEM involvement) "

THIS EXERCISE TESTED COMMUNICATIONS
INSIDE THE DAMAGED AREA BY RADIO TO RADIO HOPPING

Only radio working

Internet working

damaged area

normal area

[Drawing: G. Gibby]

NOTE: Requiring RADIO-ONLY is a **much stiffer test of winlink digital throughput!**

RESULTS:

There were 102 check-ins to their HF SSB net

"• Because most pre-designated "Official Relay Stations" did not participate, K7COW asked for relays as provided by other good signal stations on frequency like: N7CCO [official relay station], KD7SLY, K7EUG, W7PLK, and KD7JG.
• K7COW functioning as NCS is providing the only log sheets to be provided for this 75m net.
• 75m band propagation conditions generally deteriorated during the 10:00AM to 12:00PM net time period.
• It is believed most, if not all, of the stations trying to check-in were ultimately successful during the net (with the assistance of relays to K7COW).
• W7OEM's ability to effectively serve as NCS on a statewide basis is challenged probably due to marginal antennas, limited transmitter power output, limited radio operator staff resources, and other tasks needing to be concurrently performed (like Winlink communications).
• Having a pre-defined cadre of geographically dispersed stations capable to function as NCS or "Official Relay Stations" is probably a best practice.
• This SET only utilized 75m net operations. Having concurrent 40m net operations might provide better communications and propagation options and possibly reduce station check-in congestion.
• All-in-all I was impressed with the tactical professionalism, cooperation, and helpfulness demonstrated amongst the net's radio station participants (no one: asked for signal reports, wanted to rag chew, asked for SET information, or criticized what was going on). "

Very difficult to hear a lot of stations

This is from Klamath County and is what the SET designers had hoped for from relay stations spaced throughout Oregon:

:

Call	County	Heard		
		Not	Barely	Clear
K7COW				X
K7CVO	Benton		X	
KC7COL	Columbia		X	
K7HWY	La Pine		X	
AE7RB			X	
W7OW	Multi		X	
W7OEM	Salem		X	
W6OEM	Klamath			X
KD7TNG	Klamath			X
KD7CL	Coos			X
W7KHD	Klamath			X
KK6GXG	Klamath			X
KI7JMZ	Klamath			X
KE7YX		X		
KM7MM	Wheeler		X	
K6RMP	W Lane		X	
N7RCE	W Lane		X	
W7RAD	W Lane		X	
W7SZS	Clac	X		
W7RI	Lane	X		
WA7WDC		X		
AG7FR	Klamath			X
N2RSN	Klamath			X
N2RSI	Klamath			X
KF7RSF	Coos	X		
W7OEM	Salem		X	
N7DCD	Coos	X		
WA7HIW	(CW)			X
W7CLA	Clatsop	X		
W7QH	Benton	X		

(Selected) Comments Received:

"EXTREMELY useful **The first time that we had Real County Emergency Management participation.** We were able to demonstrate that we needed a better permanent HF antenna. **Our emergency folded dipole totally outperformed the HF wire antennas that were professionally installed.** Suspect a communications grant/upgrade in our future."

"We activated five stations within Deschutes County. **This SET was a great forcing function to connect areas & served agencies via previously unused or dormant tools & modes** as well as new working relationships. For example, we used P2P Winlink via digipeater for the first time from Sisters-Camp Sherman Rural Fire District to La Pine Health Center. **Our HF SSB worked well, but the PACTOR connections did not in spite of practice via MPS established in October.** Thank you Steven on behalf of DC ARES/Auxcomm & COARECT "

"Overall interesting and we learned a lot. (AAR in a separate note) 1. The scenario was too long to read. **2. 75 Meters was a poor choice for daytime transmissions. 3. An ICS-205 should have been distributed with the SET instructions. 4. Setting up MPS should be an overall Oregon activity separate from a SET.** 5. Some members thought there were too many activities and modes for one SET."

"31 MCARES operators participated in this SET. We had both the County EOC and Field station ARES Comms Trailer trying to get the "Sitrep" through to a Hybrid Gateway. **All gateways were extremely busy and this is our biggest challenge during a statewide SET. Many counties competing for the same few gateways.** "

"**The radio only mode was interesting and instructive but too few hybrid gateways to accommodate all the traffic. illustrates the need for more hybrid gateways!** MORE elsewhere in the AAR"

[all emphasis has been added by me]

Suggestions from me:
- *Radio only WINLINK under automated control (not peer-to-peer)   is very slow and the computers are not nearly as good as humans at recognizing marginal connections and moving on.*

- Normal WINLINK is reasonably functional giving the bandwidths and amateur capabilities -- no where near as fast as the Internet but much better at logistics tasks than voice.

- *Email is a terrible way to give short tactical information.*

- A well trained ham knows his tools and when to use each.

- EXERCISES make people get to know each other, build relationships, wring out problems, and develop much better groups and capabilities!

5 SECTION EMERGENCY COORDINATOR'S MESSAGE

by Karl Martin KG4HBN

Photo: Weather.gov

We will be looking over the after-action report published after hurricane Michael. The report was made to review plans and actions during Hurricane Michael. I will be covering what worked, what didn't work and what needs changing.

After reviewing the after-action report, I will be discussing what has been done in the past four months to improve existing plans, what plans will be completed by June 1, 2019, and what strategies are being made for the future.

I will take questions and comments regarding Hurricane Michael, plans for the North Florida Sections and how the Florida Tri-Sections are working together with the State Emergency Operations Center to improve plans for the next Hurricane.

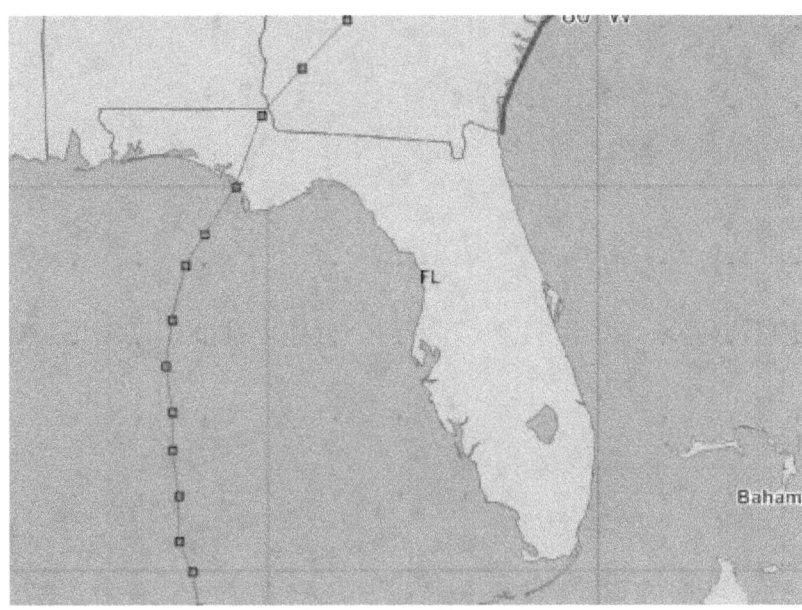

Photo: Weather.gov

Excerpt taken from North Florida Section After-Action Report.

There is no way you can ultimately be ready for a disaster that affects a large area like Michael. You can prepare, plan and simulate such an event but every deployment is different and being flexible is critical. The movement, intensity and how quickly it intensified Caught everyone off guard. In less than 18 hours Michael went from a category 2 to a category 4.

Hurricane Michael was a learning experience for everyone. Most plans in place worked as expected, some needed to be adapted to work better, but others need to be changed or updated. The operators that participated during Michael did a fantastic job. People from all across Florida came out to help. People lived a few streets distance, but others lived hundreds of miles away. This response was a team effort, and without everyone's help, we could not have accomplished what we did for Florida and our communities.

The North Florida Section was tasked to cover over 30 shelters, County EOC's and the State EOC. All of these locations were covered except for a few shelters and EOC's that felt operators were unnecessary.

Please save your questions on the next page for the Q&A session of the presentation.

Notes

6   BAPTIST DISASTER RELIEF SERVICES FOR AMATEUR RADIO VOLUNTEERS

by Marvin Corbin
Logistics / Field Missionary
Florida Disaster Relief Operations

Philosophy

The Disaster Relief Ministry allows Florida Baptist to act immediately and effectively as Jesus did to help people. When people are hurting because of a disaster, we respond with love, healing, and help. We assist all people in crisis regardless of ethnic background or religious beliefs. We do not ask the survivors for financial support.

Motivation

- Disaster Relief efforts can be summed up in one phrase: "A cup of cold water in Jesus name"

- We are following the example of Jesus when He fed 5,000 and 4,000 and when he stopped to help just one person in need.

- His teachings in the parable of the Good Samaritan- Luke 10:30-34

- Disaster Relief is Christian Love in action, meeting urgent needs of hurting humanity in crisis situations.

- Jesus often used teaching sessions to heal or healing to teach. When He sent the twelve disciples, He instructed them to teach and to heal, giving first priority to teaching. However, when He sent forth the seventy, He instructed them to heal then to teach.

Mission Statement

The mission of Florida Baptist Disaster Relief and Recovery Ministries is to "Make a difference" in times of disaster by connecting Florida Baptist Churches and Associations to people and communities impacted by disaster and by responding with Help, Healing, and Hope, that demonstrates and shares the love of Christ.

Core Values

- Christ Centered: The Florida Baptist Disaster Relief (FLDR) ministry is composed of people who are believers in Jesus Christ and have a mandate from God to minister to people and demonstrate God's love as we share the message of hope in Jesus

- Partnership Oriented: We seek to be a valuable partner, building strong biblical relationships that empowers volunteers to serve communities affected by disaster with passion, integrity, professionalism, and credibility.

- Fluid in Practice: Whatever it takes, we are His hands and feet. FLDR is known both nationally and internationally as being able to make whatever adjustments are necessary to accomplish the ministry task at hand. This is a direct result of volunteers putting others first and following the direction of our leadership team(s).

- Effective in Action: We continually seek to empower leaders, volunteers and partners through training, mentoring, accountability, and structure; enabling them to use their skills, talents, abilities and resources to be effective followers of Christ ministering to those affected by disaster.

- Local Church Focused: It is our goal to maintain a strategic and intentional partnership with local churches to support, reinforce, and encourage their ministry of making disciples of Jesus Christ and growing the Kingdom of God.

- Personal Information Form: It is important that you fill out a new PIF each time you attend training as well as each time you arrive on scene responding to a disaster.

Our History

Southern Baptist Disaster Relief (SBC DR) is a network of state convention disaster ministries that work together in large disasters. Founded in 1967 SBC DR has become one of the largest disaster volunteer organizations in the U.S.

Florida Baptist Disaster Relief (FLDR) was begun in 1982 by the Florida Baptist Convention's Brotherhood Department. In 2006 Disaster Relief & Recovery became a stand alone department responsible for all training, activation, and operational coordination of the ministry of Florida Baptist Churches.

In the wake of a disaster having trained volunteers helps Florida Baptist Disaster Relief be the go to faith based volunteer organization. Every year FLDR plans training opportunities in every region. We try our best to have training events within a two hour driving distance in each region.

Being a trained volunteer allows you to be a credentialed volunteer. Every volunteer must be credentialed to be allowed in a disaster area. During the regional training you will receive your yellow hat, shirt, and official FLDR badge. Beginning in 2015 volunteers will go through a background check before being credentialed. Background Check cost is $17.

2019 Training Promotional Material

2019 Training Locations

March 9 – Region 2 –City Church | 3215 Sessions Road, Tallahassee, FL 32303

March 23 – Region 7 –SE Region Training Ctr | 140 E. 7th Street, Hialeah, 33010

April 6 – Region 6 –McGregor Baptist Church | 3750 Colonial Blvd., Ft Myers, 33966

April 27 – Region 4 –Northridge Church | 2250 State Road 17 S., Haines City, 33844

May 4 – Region 5 –Trinity Baptist Church | 1022 S Orange Blossom Trail, Apopka, 32703

May 18 – Region 1 –Hiland Park Baptist | 2611 N. Hwy 231, Panama City, 32405

June 1 – Region 3 –North Jacksonville Baptist Church | 8531 N. Main Street, Jacksonville, 32218

First Time Volunteers

Registration Fee $25

First Time Volunteers will participate in our New Volunteer Class that is designed to prepare volunteers to be able to respond effectively with our ministry. Information will be shared to help answer the questions that all new volunteers typically have. The class lasts all morning.

After lunch, first time volunteers will be able to choose one to attend a **Ministry Area Class** to get specialized training in the area they most likely want to work.

You can cross train in other Ministry Areas by attending another regional training event.

*Price includes: lunch, training materials, ID badge, t-shirt, hat and pin. These items will be mailed to you after we have received your background check clearance. Cost for background checks is additional.

7 WIFI SHELTER BULLETIN SYSTEM TO KEEP SHELTER RESIDENTS INFORMED

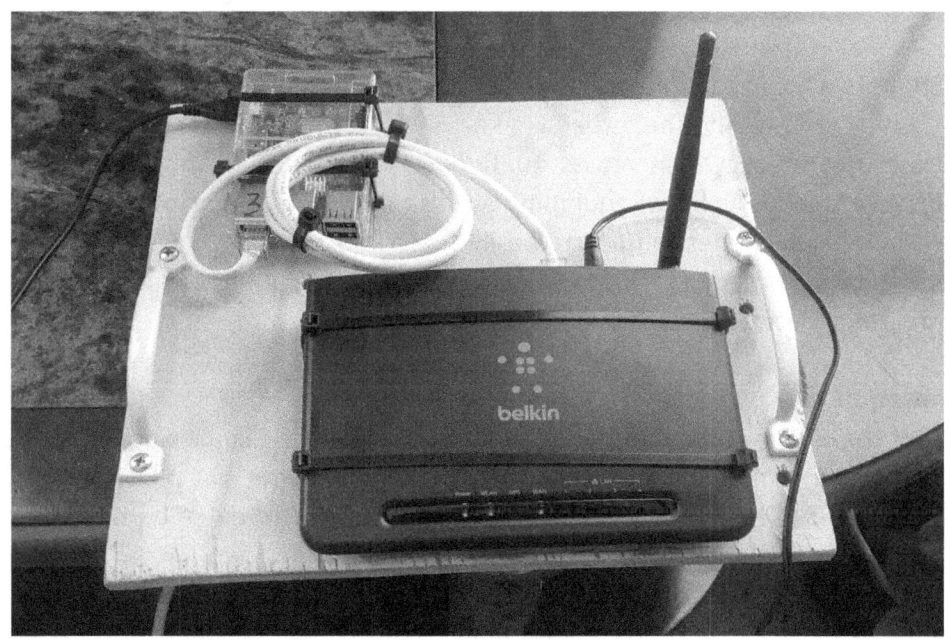

Photo: G. Gibby

This simple system worked so well that we now use it for all the slide shows at our ARES meetings.

by Gordon Gibby KX4Z

So you have 450 people -- families, elderly, small children -- crammed into the gymnasium and every available classroom of a Shelter.   All their precious cellphones seem useless --- there are no usable cell signals, there is no Internet, no WIFI......and how do you get news and other important bulletins to them? How do you show them information on how to sign up for aid, or possibilities of assistance?

We found a simple potential solution to this problem, and it costs less than $50 per shelter and maximizes the ability of HAM RADIO to help the public in time of need --- all because nearly every family these days has one or more SMART PHONES!

How This Started—The Problem Statement

Local ARES/GARS hams who served as backup radio comms volunteers during Hurricane IRMA were invited to the EOC After-Action Review. I was surprised and intrigued when the head of the shelter services group pointed out their huge difficulty with simply getting information and announcements out to the thousands of people huddled in their hurriedly-deployed shelters --- over a dozen shelters. Families were bedding down in classroom after classroom in schools and other buildings all over town, and over-worked shelter managers and helper just didn't have a very good way to disseminate information.

On Feb 1, 2018, the official Alachua County After Action Report and Improvement Plan was released (http://qsl.net/nf4rc/2018/AlachuaCountyHurricaneIrmaAfterActionReport.pdf) and included 3 improvement plan requirements related to this:

SHLT-007	Shelters were in a situational awareness isle, limiting the information know to the shelter staff and the shelterees.
SHLT-007.1	Develop a strategy to increase situational awareness to shelters
SHLT-007.2	Explore the possibility of equipping each shelter with a smartphone to increase situational awareness and more easily share information.
SHLT-007.3	Increase the amount of information shared with the amateur radio operators to ensure greater situational awareness at the shelters.

Conceptual Solution

Over half of the truly historic number of shelters opened by Alachua County during Hurricane IRMA were staffed with a ham radio volunteer. **Thus information flow was possible even if power, cell, and internet were cut off**. The EOC was staffed with a very capable crew of ham radio operators who interface well with EOC officials.

How to move information from the EOC / Ham radio volunteers to the hundreds of citizens in each shelter? In retrospect, the answer seems simple: almost every resident has a CELL PHONE and even in the loss of power and cell towers, these can connect to a WIFI router and access information – if we had a simple web server populated with important information for a storm shelter, and updated with weather information and press releases from the EOC.

An elegant solution to this soon emerged, with an inexpensive Raspberry Pi providing an Apache Web Server to any cheap consumer WIFI router. The Raspberry Pi is easy to set up so that it acts like an Internet Service Provider, dispensing all the dhcp ip numbers, nameservices and web server output required to feed information to any shelteree.

The consumer WIFI router is given a self-explanatory SSID similar to "SHELTER-A." Selecting it with a smartphone or computer gives the occupant immediate access via the easy to explain URL **10.10.10.10** That's a legal private network number than ANYONE can use without any repercussions.

A very large area can be covered by using wifi extenders, or simply by plugging in additional WIFI routers (using different WIFI channels)! The Raspberry Pi is configured to handle dozens of wifi routers, and will dutifully issue each of them an IP number. The wifi routers then use network address translation to provide the well-known private 192.168.1.X numbers to their users.

This system was quickly put together using a free Apache web server, free DNS and DHCP server (`dnsmasq`) and free firewall and security solutions. It works great! (Pages of technical details can be accessed here:
http://arrl-nfl.org/wp-content/uploads/2018/03/ShelterWebServerTechnicalDetails.pdf) The micro-SD can can easily be copied for other shelters, or for other groups to use (see later).

**Alachua County Amateur Radio Emergency Service &
Gainesville Amateur Radio Society
Shelter Web News & Information**

EOC Information

Amateur Radio Information

Screen Capture: G. Gibby

CONTENT

As originally envisioned, an opening screen allows the user to select official announcements and information from the EOC, or to access the "ham radio" pages.

Official announcements from the EOC can include weather updates, road information, fuel availability information, damage reports, information on how to apply for aid, debris pickup schedules, and other governmental information of importance to people sitting out a storm. Much of this can be "pre-loaded" into the Raspberry Pi's (and chips can be replicated so that all the servers have the same information).

We do **not** have this set up to attempt to pass the real internet along to the shelter residents (while it is still available). They would pick that up from their cell provider or from a capable school-based wifi system. Our little Raspberry doesn't have the horsepower to handle hundreds of bandwidth-hungry movie-watchers!

However, it should be able to auto-update its "official" content using cron-based ftp captures for as long as the Internet holds up.   The EOC would need to make content available for capture for this purpose.

When the Internet is GONE is when the hams come into active play.   Updating of the system with official bulletins is quite possible because our ham group has the ability to transmit computer files faultlessly by not one, but several methods!

1. **WINLINK** provides a way to send email to all the shelter volunteers including attachments, and is error-free.
2. **Packet radio** provides the error-free file transfer protocol YAPP that even accepts files in an unattended ham radio station if so configured.
3. The NBEMS **FLDIGI**-based system also includes error-corrected methods for file transfers.

Once the update weather and other announcements are received from the EOC, it is easy to transfer them (over WIFI) to the little wifi server, using either a.. .bat batch file, or free point-and-click FTP client software such as Coffee Cup FTP (https://www.coffeecup.com/free-ftp/).   The little Raspberry Pi has all the passwords and firewall and other security to make it a secure and reliable system.

HAM RADIO CONTENT

Shelter residents quickly become BORED.   On the amateur radio page, store all the ARES, radiogram, procedures and instructions you wish, but also put some entertaining & educational information on ham radio, how to get into the hobby, and maybe even put in some tutorials to get people started toward getting their first license!! The little Raspberry Pi server probably can't handle streaming many videos, but you could easily have several simple educational slide shows etc.   One of our members is exploring a way to download the entire Wikipedia and serve it!

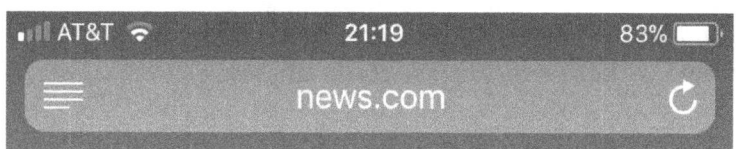

ALACHUA COUNTY ARES ≈ GAINESVILLE AMATEUR RADIO SOCIETY

Volunteers from both these wonderful, local amateur radio groups staff shelters in Alachua County during times of stress. Amateur Radio is a fascinating hobby, with over 700,000 Americans and hundreds of thousands of others all over the world enjoying the diversity of this hobby. Hams may build and operate their own radios, experiment with satellite communications, participate in emergency services, or just meet friends over the radio. In the United States, all are licensed by the Federal Communications Commission (FCC) after passing multiple choice tests that range from fairy easy (for the introductory Technician license) to rather complicated (the Amateur Extra Class license).

This hobby is built on volunteers! The Gator Amateur Radio Club, at the University of Florida, as well as ARES, conduct periodic classes to assist people in passing the license exams, and another group of hard-working volunteers provide at-cost examinations every other month in Gainesville.

Below are links of interest to persons curious about ham radio, and also stored information to assist our volunteers staffing shelters:

Alachua County DOT Map
Alachua County Attractions
Amateur Radio FCC Band Limits
Alachua County (Draft) Comm Plan
ARES Field Manual
ARES 2015 Manual
Cascadia Rising FEMA Exercise
COMT Pre Course Study Guide
EasyTerm Tutorial
FLDIGI Cheatsheet
Georgia Hospital Emergency Comm Plan
Katrina--Report
Marion County Hospital Emergency Comms Plan
NIFOG
North Florida Section Comm Plan
Radio Relay International Traffic Manual
Simplex Repeater User Manual
Soundmodem Configuration Information
SSTV Manual
Sound Card Interface Full Manual
TrafficNets List
WINLINK - Cheatsheet
WINLINK HF Primer
WINLINK without WINLINK SOFTWARE

How To Implement This In Your County

If you are reading this at our 2019 EMERGENCY COMMUNICATIONS CONFERENCE -- you can have a card burned for you right here -- or you can download the 3 Gbyte zipped image file and copy it onto a micro SD card of your own, either now or later.   Ask for the details!

If you would like information on how to get a free working copy of the entire Raspberry-Pi software, contact the author at docvacuumtubes@gmail.com.   I can copy onto your microSD card. Simply plug into any Raspberry Pi version 3, and then connect a wifi router.   As you'd expect, there are no guarantees or warranties and the author cannot accept responsibility for any untoward events, and you will have to learn a bit about transferring files over ham radio, but this looks like a very useful solution that can help amateurs better serve their communities.

8   INTRODUCTION TO PUBLICITY / PIO

Scott Roberts KK4ECR, Section PIO

The following article was initially published in the June 2018 edition of the CQ Magazine.
http://www.cq-amateur-radio.com and used here with permission of the authors and publisher of
CQ Magazine. Please don't copy elsewhere without permission of the publisher of CQ Magazine.

The Successful PIO
Scott Roberts KK4ECR, Clay County ARES AEC/PIO, Joe Bassett W1WCN, Clay County ARES AEC

Success in real estate is about "location, location, location." Success in public relations is about...well, it's right there in the name, "relationship, relationship, relationship." Successful Public Information Officers build good relationships with public officials, journalists, news producers, meteorologists and community leaders. The danger, however, is in treating these relationships as commodities. These contacts are people, just like us, people with interests, concerns, aspirations, dreams and loved ones in our local communities. And the adage is true, "they don't care what we know, until they know that we care" about the community, too.

Relationships start with a foundation. The foundation for a PIO is their investment in the community. Attend community events, volunteer to be an election poll worker, participate with local charities, join Rotary or Kiwanis. In short, create opportunities to know and be known.

The foundation for any relationship is the first impression. The people that you meet in community settings are looking for social benefits), other people that add to the quality of life. These benefits can be any of four social gifts: appreciation, connection, elevation, and enlightenment (as defined by Demarais and White in the book First Impressions). As you prepare to make a first impression with community leaders and the media, think through the above list. Appreciation: Is our organization thankful for the opportunity to serve the community or do we begrudge others who might not understand our value? Or elevation. Are we condescending in our interactions with other agencies or do we seek to help everyone improve, including ourselves? How about connection? Are our community relationships based on mutual respect and appreciation?

These are important points to ponder as we prepare our ninety second "elevator speech." The more the PIO is involved in the community, the more opportunities to introduce ourselves will produce themselves. When that opportunity strikes, be ready to seize it by presenting those four gifts. **Appreciation:** "We Amateur Radio operators are thankful to have partnered with relief efforts during the recent hurricane." **Connection:** "Several of our members are active in your Rotary club. Did you know that the treasurer is an Amateur Radio operator?" **Elevation:** "Our Amateur Radio group supports a local school radio club that reinforces the STEM curriculum in science and provides Boy Scout merit badge counselors for the local council." **Enlightenment:** "Many people are interested to know that Amateur Radio isn't dependent on infrastructure in a disaster. When other communication systems fail, we can usually get messages through."

Once the successful PIO makes that first impression, it's imperative to nurture it. Share your card with those that you meet and make sure to ask for their contact info. Assure the new connection that you'll be happy to contact them if you think of future opportunities where Amateur Radio might benefit their organization or community. Then, follow up the first impression with a quick email or note thanking them for the chance to meet. This follow up is particularly important with the media. Press releases are an excellent vehicle for announcing upcoming events or highlighting service to the community, but they'll provide more return on investment if the recipient recognizes the name of the PIO. When sending a press release, precede it with a short personal text or email, "Please be on the lookout for a press release I just sent. You came to mind as someone who might be interested."

A short thought about press releases, people in the print and broadcast media are jealous of their time, and rightly so. In spite of our personal bias, most of them are honestly trying to provide a service to the local community, just like us hams. If you "write the story" for them in the press release, they will be much more receptive. Make sure to include the what, when, where and why. The why is the most integral. Why is this important?

Once the successful PIO has made a first impression and followed it up, he or she will maintain the relationship. Keep in touch with your contacts. Let them know when and where radio operators will be active and contributing to the community. A good way to do this is with "personalized bulk emails." In Clay Ares we use Google Sheets with the Mail Merge add-on. People are proud of their name and personal messages go a long way to improving relationships.

In 2016, Clay Ares sent press releases announcing Field Day to each broadcast media outlet and more than 75% of the print media in northeast Florida. A few cursory phone calls of interest were received and there was no media coverage of the event.

In 2017, we sent press releases *and* personal invites/notices to every person on the list we had cultivated with the above principles. The return on the investment was an appearance on a local noon-time talk show, several government officials visited our operations and invitations to speak at all county

high schools were received.

Then, when hurricane Irma was barreling toward Florida, the media was seeking us out. As word of our preparation to support the local community spread, 46 inquiries resulted from press releases and one prominent reporter attended our monthly meeting immediately preceding the hurricane. He stayed for the entire meeting and then produced a news package for that evening's news. In the last 12 months, Clay Ares has been featured on five different news broadcasts.

The lesson we learned? The successful PIO doesn't wait for a disaster, expecting the public accidentally realize that Amateur Radio is ready to help. The successful PIO, and the entire organization, continuously builds relationships with the community; relationships that benefit society, not just the hobby.

As stated before, the foundation for building any team is "relationship, Relationship, RELATIONSHIP!"

We need to remember that each person joins the Amateur Radio hobby for different reasons. For some, they got their license to talk around the world. For some, it is because they wanted to use the computer over the radio. For others, it was to help the community in times of emergency. And then for others, they did it because grandpa made it look like fun.

As we build our teams it is our job to find out why people became Amateur Radio Operators and then build on their "WHY."

So, take a moment to ask yourself, "Why did I become an amateur Radio Operator?"

With the potential for so many different, "WHYs" how do you successfully build a team that everyone enjoys?

We have to remember that...

...it is not ONLY about...	...but, it is about...
Passing traffic,	Passing traffic.
Checking into nets,	Checking into nets.
D-Star, DMR, NXDN, Fusion,	D-Star, DMR, NXDN, Fusion.
PSK-31, RTTT, Winlink,	PSK-31, RTTT, Winlink.
Exercises in the field,	Exercised in the field.
Building antennas,	Building antennas.
Emergency operations,	Emergency Operations.
Operating QRP,	Operating QRP.
...etc.	...etc.

Amateur Radio is about ALL of these. Whether it is an Amateur Radio Club or an ARES/RACES organization, we have to have a balance of all of these things to appeal to different type of people. We have to remember that it is a HOBBY, and we need to make sure we do not take the FUN out of the HOBBY.

If we want to build a team that maximizes ALL of our volunteers, then we need to build a team that maximizes the interests of our volunteers. We have to get to know the people who are on our team and what they enjoy and what their strengths are.

Without the foundation of "Relationship," we can not begin to build a team that has <u>maximum</u> potential of being a **great** team.

9 MOVING TRAFFIC AND TRAINING VOLUNTEERS IN ARES NETS

by Joe Bassett, Jr. W1WCN

Effective emergency communication is about one thing: communication. It starts with communication. Communication is at its center. Ultimately, it's about communication. This is why the word "communication" is in the term itself. Furthermore, communication of no value if it isn't effective. Ineffective communication is a *non sequitur*, it's noise.

Related to amateur radio emergency communication, radio waves should be utilized to communicate information that results in an effect, i.e. necessary action initiated via a Skywarn report, deployment of needed supplies in the wake of a crisis, comfort in the form of a Safe-and-Well message. Whatever the message, via whatever mode, emergency communication should render an effect.

In that light, amateur radio operators are most effective as communicators who happen to use radios, rather than radio operators who happen to communicate. Communication is the goal, radios are the tools.

The decision to use a radio to send a message is too often answered with the question, "Can I send this by radio?" But effective communicators ask themselves, "*Should* this be sent by radio?" followed by, "Is radio the best means to convey this information?"

Of course, amateur radio has prided itself on the premise of "When all else fails…" Certainly, if all other forms of communication have failed, then use radio waves is warranted. At that point, radio is the appropriate vehicle and amateur radio operators are often the "cavalry" saving the day!

And, yes, there is still the possibility, probability, even eventuality, that all else *will* fail. This is a lesson learned through the loss of 911 capability in local municipalities *and* in large scale disasters such as hurricane Maria in Puerto Rico.

Here in lies the tension between the amateur radio operator's affinity for radio technology, the commitment to serve society, and the pride of proficiency. Effective amateur radio operators balance love for the hobby, drive to proficiency, and a desire to be part of something bigger than themselves.

First, love for the hobby is best revealed in the confession that we hams, if we're honest with ourselves, are still big kids playing with walkie-talkies in the woods. It's imperative to remember this while recruiting and training amateur operators toward emergency communication. Picking up a microphone and speaking information to a listener at the other end is the most intuitive form of radio communication.

Second, many amateur radio communication groups lament a difficulty in recruiting and assimilating new members. At the same time, there is a growing insistence that digital modes of communication be the foundation for supporting disaster response. It is true that digital communication is accurate, however it may not be as effective for the new ham. To illustrate this, we can ask the following question of ourselves...
"How many of us began the hobby of amateur radio to talk on the radio versus how many of us began it to type on a computer?"

It's safe to say that the former is true rather than the latter. It's no wonder that the attrition rate of "newbies" is high in many areas. The new ham is most likely attracted by personal interaction and a desire to contribute to a team, many while monitoring emergency amateur communication as scanner enthusiasts. Then we initiate them into service by insisting that they expend additional capital, time and financial, toward impersonal digital modes, before they understand the underlying principles of emergency communication. (Look no further than the lament surrounding the rise of impersonal interaction brought about by texting and social media.)

Third, as stated earlier, digital communication is accurate, but it is less robust. This is evidenced by the fact that it introduces more failure points at both the transmission and reception location. These failure points include both equipment and personnel. When additional equipment is introduced to a signal flow, potential mechanical and electrical failure is increased. Also, additional equipment requires additional training and documentation.

This is not to say that digital modes don't add significant value to emergency communication. To say so negates the truism that amateur radio emergency operators are communicators first and radio operators second. This author has stated publicly that "...amateur radio operators who neglect digital modes as valuable tools for emergency communication do so at the peril of dereliction of their duty during emergency situations." The point is: digital modes are applicable and effective after the basics of radio communication are understood and practiced. These basics are best learned through the practical application of voice communication.

In the case of digital communication, when a point of failure raises its head, *and be assured that it will*, the skillset of voice communication will acquit itself admirably. With that, amateur radio communication begins and ends with passing messages via voice. By extension, so should training.

Such training hinges on three interdependent components: directed-tactical net procedures, message handling (transmission and reception), and message formatting. In turn, each component rests on interdependent skills and disciplines.

The application of these disciplines provides an environment conducive to effective message transmission as does digital communication. In many cases, voice messages can be transmitted and received/acknowledged with more expediency than digital modes. However, there are myriad instances when digital communication is preferable to voice. These include, but are not limited to: lengthy forms, messages with no time value, and recurring messages. Additionally, many digital modes of radio communication succeed in conditions which render voice communication inoperable.

Once the voice skillset is assimilated by the amateur operator it is easy to apply that skillset to digital modes of radio communication. By extension, the understanding of voice procedures undergirds digital communication. Thus, many options for success are available to the well-rounded and equipped amateur radio operator when all else fails.

On a personal level, the author has found several beneficial maxims for training amateur radio emergency communicators...

- Practice doesn't make perfect; perfect practice makes perfect.
- If you think that you're going too slow, slow down.
- When crisis happens, do the best that you can. You can improve later.

In practice, there are two forms of messages passed through emergency or disaster response nets: 1) tactical messages, 2) record traffic.

Recent events such as hurricanes Matthew, Irma, Florence, Michael, and most famously Maria, reveal that tactical messages comprise the bulk of information passed via amateur radio in the immediate aftermath of a disaster. These tactical messages can be either written recorded messages or simply voice transmissions of information between locations associated with served agencies, such as hospitals or shelters.

Record traffic, on the other hand, is written messages produced by the served agency or individual. This traffic becomes part of the public, official record of the event and can be comprised, but not limited to, either ICS-213 messages or ARRL Radiograms. In the case of ICS-213s, the information is commonly logistical requests or routine status reports. Radiograms most often contain Safe-and-Well information.

While it is true that contemporary modes of communication such as landline, cell, and internet are susceptible to outages during catastrophic events, it is also true that such outages are relatively short lived. Even in cataclysmic events such as hurricanes Maria and Michael, more than fifty percent cell coverage was restored within seven to ten days after landfall. In the aftermath of hurricane Maria, which damaged ninety-two percent of Puerto Rico's cell towers, some areas experienced at least sporadic restoration within the first week.

With the advent of Cell on Wheel (COW) and Cell on Light Truck (COLT) technology, the length of service disruption following disasters will continue to decrease. In essence, by the time that routine Safe-and-Well traffic is appropriate or possible, messages of that nature will be possible by means other than radio frequency.

On the other hand, the ability to transmit messages of a tactical nature and which support first response is imperative in the first hours and days following a disaster. This demands and ability to respond quickly, in the first hours following the disaster. It also requires skills which have been honed and equipment prepared in advance of the event. Not just days or weeks in advance, but months or years ahead.

These initial communication needs are best met through tactical nets. The effectiveness of both directed and tactical nets rely on each station knowing and adherence to specific procedures. The general parameters and general guidelines of a specific net are outlined in that net's preamble. However, stations unfamiliar with that net's operation should refrain from participating, unless compelling conditions apply.

For the purposes of this discussion, tactical net procedures will be addressed. Please keep in mind that all tactical nets are directed nets, but not all directed nets are tactical. Therefore, the procedures discussed below are easily applied to other directed nets such as traffic nets.

As the name implies, the net control operator is responsible for control of the net. It is inappropriate for stations on the net to transmit without express permission from the net control station, such as requesting check-ins or directing communication. In the same vein, stations should follow the net control station's instruction's in form and function.

Stations participating in tactical nets are assigned tactical call signs at the discretion of the net control station. Tactical call signs might be determined by location, function or other criteria. For example, a shelter located at the emergency operations center might be assigned the tactical call sign "EOC." This call sign allows easy identification of the stations role and location. Also, the call sign can be used by the multiple stations which operate from that location throughout an extended event. Stations operating under a tactical call sign are required identify with their FCC issued call sign in compliance with FCC Regulations Part 97.

A useful acronym for effective communication is A (accurate), B (brief), C (complete). Use of procedural words (prowords), break tags and phonetics supports this ideal and is invaluable in effective tactical nets, particularly when operating simplex between stations. These streamline the process of checking in, listing traffic, and conducting communication.

When communicating tactical information, that which is not recorded or written, transmitting stations should use as few words as possible, while still conveying the full intent of the information. These transmissions may be requests for timely information, notification of status changes or similar. Tactical communication might or might not be recorded in the log; this is at the discretion of the stations communicating. Also, tactical communication is most often generated by the radio operator.

Record traffic, on the other hand, is always logged by all stations participating in its transmission and reception: sending station, receiving station, relay station, and net control station. Record traffic is most often generated by the served agency and should always be transmitted without modification by the radio operators involved.

Record traffic is introduced into the net either at check-in or by establishing communication with net control. If the originating station introduces the traffic, then the proword LISTING is used. If the message is not introduced by the originating station, then the proword HOLDING is used. Below is an example script of the former:

NET CONTROL this is SHELTER 1. I LIST one ROUTINE ICS-213 for EOC.
SHELTER 1 this is NET CONTROL, ROGER, OUT.

When introducing traffic into a net, three pieces of information are required:
- PRECEDENCE (routine, priority, or emergency)
- TYPE OF FORM (this informs the receiving station which form to have at the ready)
- DESTINATION (this informs Net Control as to which station to receive the message)

Regarding type of form, most record traffic is proforma in nature. That means that the form will always follow the same structure and flow. This allows messages to be transmitted without repeating each preamble or section heading. When multiple forms are being used, enumerating which form contains the information allows the receiving station to be prepared.

As to destination, most often the destination will be another station on the net. In the event the destination is outside the purview of the net, then the following prowords would be used:
- Outside the immediate county: NET CONTROL this is SHELTER 1. I LIST one ROUTINE ICS-213 for OUT OF COUNTY.
- Outside of state: NET CONTROL this is SHELTER 1. I LIST one ROUTINE ICS-213 for OUT OF COUNTY.

By listing the traffic in this manner, the net control station can route the message through the most effective station.

When the net control station is ready for the message to be transmitted, the interchange would be similar to this:

SHELTER 1 this is NET CONTROL, call EOC and pass your one ROUTINE ICS-213, OVER.
NET CONTROL this is SHELTER 1, WILCO, OUT.
EOC this is SHELTER 1, OVER.
This is EOC, OVER.
1 ROUTINE ICS-213 for your station, OVER.
ROGER, OVER.
MESSAGE FOLLOWS [transmits message]

Following the transmission of the message, the receiving station would either acknowledge receipt with the proword ROGER, identified with the serial number of the message, or request fills.

Notice that SHELTER 1 did not begin passing the message until establishing communication with the EOC and receiving confirmation that the EOC knew which form to use. Also, transmitting stations should not break for fills until the entire message has been transmitted.

This is by no means an in-depth discussion of tactical net procedures and traffic handling. Many other disciplines are involved, such as providing fills, distributed frequency utilization, digital transitions, etc. It is simply an introduction to the best practices of directed net discipline. Readers are encouraged to seek out further resources provided by the ARRL and myriad of online resources. Also, simply reading about these procedures is insufficient for effective radio communication. Practical application in simulated events is imperative.

10   COMPUTER AND INTERNET TIPS FOR EMCOMM

by Jeff Capehart W4UFL

Computer & Internet Tips for EMCOMM
PART ONE

Computers have become an integral part of everyday business and that applies equally to emergency communications. Whether reading email, looking something up on the Internet, editing a document, or printing a PDF file, we all need some basic skills in using computers to get things done in the digital age. Those who work on a computer every day will have a slight advantage as their tool skills will be more practiced. However, there are many tips and secrets to Windows that remain hidden within the computer until you unlock the knowledge of how to use them.

QUESTION 1:
How do you get pictures off your phone and onto the computer?

ANSWER: If you are emailing them to yourself, that won't help you much if the Internet is out. Use your USB charging cable and mount your phone as a disk drive to copy files!

Choose the option to open your files in a folder. If you don't get a pop-up window for your flash drive, open **Windows Explorer** and look for a 'new' device typically on the E: drive or G: drive. Be sure to scan all the files on the device with your anti-virus software if it is the first time putting it in the computer and especially if your anti-virus is not set to automatically scan.

You'll typically find your photos are in the DCIM folder. DCIM stands for Digital Camera Images.

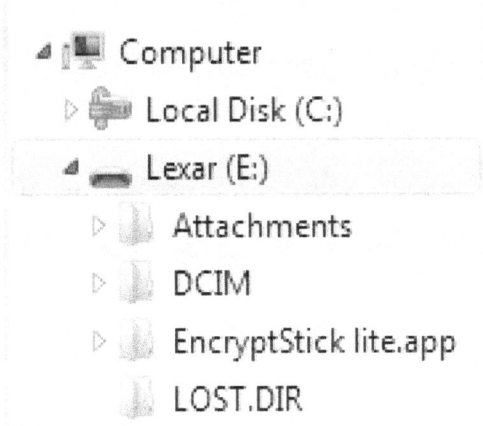

Windows Folders & Copying Files

The easiest way to copy files is to use the Folder View. Simple use the mouse to select your file or folder and drag it to the destination folder.

You can easily make new folders to organize your files by click on New Folder. Be sure to use a descriptive name that will help you remember what should go in the folder.

Windows Versions

Windows comes in many versions, but the current release is Windows 10. Older versions are either already off support or going out of support. Windows 10 has been out since 2015! Windows XP was end-of-life years ago back in 2014. The low cost of good laptops, the improved security in Windows 10, better battery technology, and use of solid-state drives have dramatically improved the usefulness of even the low-end portable computing devices.

Getting Around Windows

Most users of Windows have figured out how to use the mouse to get around. Clicking once to select. Click twice (double click) to activate or launch a program. Click and drag to move an item. That's the normal left-click button. The right click button brings up a context-sensitive menu that can be used to perform additional functions that are not always obvious.

The mouse can also be used to select text, copy text, and paste text – all using the right mouse click. The copied text goes to the 'clipboard' in Windows.

There is also the capability to copy graphics using the snipping tool in Windows 7. Built-in tools such as Microsoft Paint will allow you to paste and edit your images so that you can get just the right selection. This is extremely useful if you are creating a document such as a how-to manual and you need to show portions of screen shots.

Special keys are available to perform screenshots of what is currently displayed on the computer screen. That image is also copied to the clipboard and can be pasted.

Select All: CTRL-A
Copy: CTRL-C

Paste: CTRL-V
Cut: CTRL-X
Screenshot   PrtScr
Windowshot   Alt-PrtScr

Virus Prevention

Anti-virus software has been a mainstay of good security practices for over 25 years. New computers generally come with some type of commercial Anti-Virus software on a trial basis that you then have to pay for after the trial period expires.

Microsoft offers Windows Defender, a free virus checker program that is pretty good. For Windows 10, Windows Defender Antivirus is built-in. There's nothing to buy and nothing to install. No configuration, no subscriptions, and no nagware.
https://www.microsoft.com/en-us/windows/windows-defender/

Some ham radio software requires overriding the virus checking. Usually this is because the software author did not get registered to be able to code-sign the software. Check with other hams to determine if the software is safe if you have questions about its authenticity or use.   And be sure to always download direct from the software author's website.

Installing Applications

Windows comes with a lot of stuff built-in, but not always the kinds of programs that you will need. The corporate world can afford to put the Microsoft Office suite of programs on their knowledge workers' desktops. Word, Excel, PowerPoint, and possibly even Adobe Acrobat. Not the free reader, but the version that allows you to make and edit PDF documents.

FREE Office Productivity Software is available. OpenOffice and LibreOffice are two software installations that you can use at no cost. OpenOffice does not appear to have a Windows 10 edition, so many of our volunteers have started using LibreOffice.
https://www.libreoffice.org/download/download/
https://www.openoffice.org/product/windows.html

One key item of note to mention is that you will need to know what version of operating system you are using and whether it is a 32-bit or 64-bit edition. If you download the wrong version, it may not work, or it may not work very well.

QUESTION 2:
Who knows what the "Windows Key" is on the keyboard and what it is used for?

ANSWER:
The Windows Key is used for shortcuts to commands and functions, but you have to memorize them or keep a cheat sheet handy!

Example: Pressing the Windows key and the "R" key will bring up a dialog box to run a command. Type "msinfo32" in the Run dialog box, and click on the OK button. Try this to find out your computer's "System Information".

FREE Ham Radio Software is also available for your computer. Programs like Winlink, FLdigi, Soundmodem, and Easyterm will provide you with an array of programs that can send email over ham radio, transfer files, do packet, and do many of the popular digital modes like PSK31.

Download Site Links:

Winlink Express

Winlink Express (current production version)   https://winlink.org/WinlinkExpress

https://downloads.winlink.org/User%20Programs/Winlink_Express_install_1-5-17-0.zip

Soundcard Modem Driver

For VHF packet, you can download a soundcard interface and a modem program. We have been using UZ7HO's free soundmodem.exe:

SOUNDMODEM: http://uz7.ho.ua/packetradio.htm   Latest Version 100 as of April 6, 2018

soundmodem100.zip   06-Apr-18 07:02   459.99K

Terminal Program for Packet

While you are there at UZ7HO, download his excellent plain-jane terminal program called Easyterm:

EASYTERM   http://uz7.ho.ua/apps/easyterm41.zip Latest Version 41 as of Sept. 19, 2018

easyterm41.zip   19-Sep-18 11:44   400.86K

Dealing with ZIP files

Occasionally when you download from the Internet, you will get a ZIP file. These are compressed archives that contain many files. The compression varies, but can reduce the file size to make it transfer faster over the network. Downloaded files from the Internet are saved by default to the "Downloads" folder.

Don't be afraid to download a ZIP file. These files are sometimes blocked by email because they can be large and typically are used to sneak malware past your email system's virus scanner.

Ever since Windows XP, Windows has had a native ability to open, view, and extract files from a ZIP archive. You can see the files in the archive and open them, but not every program can. So, the best thing to do is extract them from the ZIP archive to a folder where you want them to go. You can scan the ZIP file and then scan the contents after extraction just to make sure.

Windows Key & Keyboard Shortcuts

Use the Windows Key to control the ways you interact with the windows in Windows.

https://en.wikipedia.org/wiki/Windows_key

Pressing the WinKey once brings up the start menu! No mouse needed; no hunting for it.

Win + D will hide all the windows and show you your desktop.
Win + R will open up a Run Command window.
Win + E will open up Windows Explorer.

Key Windows Tips for Making Life Easier

Check your computer's storage to see how much room you have for programs and data.

Open Windows Explorer. Go to COMPUTER and click on it. The free space and total space for each disk drive is displayed, along with a bar graph representing the percentage utilized.

Demonstrations

1. Connecting to WiFi using Windows
2. How to type into a fillable PDF document
3. Saving documents to your "Documents" folder
4. Organizing your documents by creating folders (Projects, Calendar Year, Groups)

Windows Device Drivers – Finding and Solving Problems with Hardware

Windows 10 sometimes doesn't work well with older designs of hardware. This is especially true for USB-to-SERIAL interface devices for connecting up older ham radio equipment like TNC's (Terminal Node Controllers) for PACKET radio. You may also have an older radio programming software cable that needs a serial connection. Be careful of devices that use Prolific chipsets. The "good" devices use FTDI. If you have one that isn't working, check the **Device Manager** for "**Prolific USB-to-Serial** Comm Port (COM#)".

Device Manager is the Windows way of installing and managing the custom device drivers for hardware that is manufactured by third parties. There are several ways to get into the Device Manager panel and determine what is going on.

The easiest way is to just use the search box on the taskbar, type **Device Manager**, then select from the menu. An alternative way is to run the program directly, but you have to know the exact name!

1) On your keyboard, press **the Windows logo key** and **R** at the same time to invoke the run box.

2) Type **devmgmt.msc** and click **OK**.

In Device Manager, you are typically looking for yellow exclamation marks or red X marks that indicate a device is not working properly or that the driver failed to install. You can then make changes or reinstall the driver. The error codes can help you search for ways to fix your problem.

Web Browsers

There are many different browsers for the World-Wide-Web on the Internet. Microsoft's Internet Explorer has been replaced by the Edge browser. Chrome, Firefox, and Opera are other browsers that may work equally well for you. They are easily found with a search by name and usually have key features to distinguish themselves from the others. Edge is the built-in browser with Windows. Google users may prefer to have a common experience across all platforms using Chrome for their desktop, laptop, tablet, and phone. Firefox has many add-ins especially the NoScript add-in which lets you control which sites are allowed to run javascript.

**Computer &
Internet Tips for
EMCOMM
PART TWO**
Purchasing Computers
Updates
Advanced Skills

Buying a New Laptop Computer – On the cheap!

Buying computers may seem like it takes an expert. But these days, you can get a perfectly good computer for just a couple hundred bucks. But how to know what you are getting?

First, don't get a "Netbook". Those are limited computers that rely on the Internet to do all the work. They are mostly for web browsing and email. You'll want to look for particular hardware that will make your experience much improved.

CPU – The Central Processing Unit. The computer chip. Intel makes the i3, i5, and i7 processors. The 3 is the low-end, 5 is mid-grade, and 7 is high end. The i3 works well enough with sufficient computer memory (RAM) in the 4-8GB range. Find a refurbished computer and save hundreds. Use Amazon to find deals.

Lastly, think about your storage options. A solid-state drive (SSD) is a fast and reliable, but more expensive alternative to the traditional spindle-based Hard Disk Drive (HDD). Hard Drives come in 500GB – 1 TB. A terabyte is 1,000 gigabytes.

Basic Computer Security

1. Use anti-virus software

2. Install hardware and software firewalls

3. Create strong passwords

4. Establish a back-up schedule for important data

5. Maintain up-to-date security patches

6. Use password-protected screensavers

7. Check security settings in your e-mail client and web browser

8. Use safe e-mail and download practices

9. Increase your awareness of Internet security

10. http://www.secureflorida.org/

Additional Safety on the Web

Don't give away your social security number.
Be careful receiving emails that look suspicious – they are usually scams. They try to trick you into entering your username and password into a look-alike-website that fools you into thinking you are on the official site.

Microsoft Defender Anti-Virus
Microsoft has stepped up its efforts in security with Windows 10 creating improvements to Windows Defender. The Windows Security Center is now the Windows Defender Security Center.

Microsoft's Windows Security Research Team benefits from a vast installation of over 1 billion consumer versions of the antivirus engine.

Windows Defender, although built-in to Windows, typically has to be turned on and replace whatever third-party vendor anti-virus software was pre-installed with your new computer. Uninstall the third party stuff and use Windows Security Center to turn on Windows Defender.

Be careful with some portions of Winlink software. For example, the VARA protocol requires a special download and installation that gets pegged as a Trojan by some systems. If you have downloaded it from the original site, it should be fine and you can exempt that program from the anti-virus scanner and it will allow it to function properly.

Windows Update – Automatic or Manual?

Windows 10 now defaults to Automatic Updates. This is really good for security because your system will check for updates directly from Microsoft automatically. However, in some cases, there are upgrades to both software such as WINLINK and your Microsoft Windows that occur frequently. Microsoft has "Patch Tuesday" on the second Tuesday every month. Other emergency and critical security updates could be released sooner. One way to keep a computer both secure and ready-to-go is to always keep a spare copy of working INSTALLATION SOFTWARE on a USB flash drive. That way you can always re-install even if you don't have access to the Internet. If you are really concerned, you can disable Windows Update or put it in manual mode, but then of course you have to remember to do it yourself!

More Demonstrations

1. Searching in PDF documents

2. Searching on the Internet

3. Opening Multiple Windows

4. Opening Multiple Browser Tabs

5. Using Task Manager

6. Windows Key (Right Arrow / Left Arrow)

7. Credential Manager

11   WIRING YOUR RADIO FOR SIGNALINK / DIGITAL

by Gordon Gibby KX4Z

Digital wiring is not nearly as complicated as some people make it out to be. You need exactly the same connections to use a soundcard digital mode (packet, pactor, Winmor, psk31, Olivia, rtty, ARDOP or VARA) that you need to connect the same radio into a repeater, or into a phone patch. Exactly the same type connections that hams made during the Vietnam war to provide long-haul phone patches from troupes back to their family!

RADIO MICROPHONE -- has to get sound from the sound system, and it is very low signal level, usually 10-30 millivolts, and thus very susceptible to interference.   For this reason, shielded, or at least twisted pair w/ground wiring is used for this. Further, the radio mic wire may have a dc power voltage superimposed on it (for the purpose of powering an electred mic) -- so devices such as Signalink's include a series capacitor so this voltage isn't shorted to ground.

GROUND – and some radios have two different "grounds" -- one for the microphone that is separate from the "ground" for the push to talk circuit.   Most Signalink or other sound card isolation system don't have TWO ground outputs, so we just end up connecting these together.

AUDIO OUTPUT FROM THE RECEIVER – this has to go into the microphone INPUT of the sound card system   (the radio's OUTPUT is the sound card's INPUT and vice versa). This is typically at least 100 millivolts.

PUSH TO TALK -- and for modern solid state rigs, this is almost always a positive voltage which must be shorted to ground in order to cause the transceiver to go into TRANSMIT.   But not always!!! Be careful with Motorola walkie talkies -- they may use a very different system.   Much older rigs (vacuum tube rigs) may use a substantial NEGATIVE voltage. Most connection systems switch the push to talk with a RELAY for these reasons so polarity isn't a problem (but some still use a transistor or other device)

GETTING ALL THOSE SIGNALS

A very few radios (older ICOMS for example) include all the necessary signals on their microphone connector. Many radios however don't have the receiver audio output and you have to capture that with a 1/8" (3.5mm) plug into a jack on the back.   Some rigs use a very special connector on the back of the radio ("auxiliary" or similar) with a plethora of signals.

For each radio, you'll simply have to read the owner's manual to find where the signals are. Be careful of some radios where you can get the signals off the back connector --- but the Microphone is still live and will pick up room sounds if not disconnected. This can be a problem since the digital modes are typically used in the CW/DIGITAL sections of the band – not in the voice portion of the bands.

WIRING TO A SIGNALINK OR OTHER DIGITAL SYSTEM

The popular SIGNALINK allows you to internally configure which signals appear where on its RJ45 socket --- leading to problems if yours is wired differently from the other one someone loans you. For this reason our emcomm group in Alachua County standardized on one pin out, so that volunteers would always know the signal pin out on the sound interface card RJ45 plug, and as much as possible, we also use the same pin out on all Signalinks in use; it is the one recommended for popular Baofeng UV5RA radios, and also for mini-6-pin DIN radio connections.

ALACHUA COUNTY FLORIDA STANDARDIZED RJ45 PIN OUT

PIN	SIGNAL	Wire Color
1	microphone	white/orange
2	ground	orange
3	push to talk	white/green
4	unused	blue
5	receiver audio	white/blue

This table assumes the numbering shown in the accompanying Figure where the gold pins are held upwards and away from the reader.

Figure *Top View (pins visible) numbering of the RJ45 plug pins.*

This happens to be the standard pin out of commercially available cables intended to connect the popular Baofeng UV5RA (which uses a special molded double-plug connector) to a Signalink. This allows one to easily plug in a Baofeng UV5RA low power transceiver to test a system.

RFI REDUCTION

(A bit of trial and error, usually, and VERY important.)

To reduce RFI, put 3 or 4 loops (2-3" dia.) in the audio cable from the radio to your digital interface system and also clip on a ferrite "bead" if possible.

Laptop touchpads may frequently experience RFI when you touch them while transmitting. In that case, use a wireless mouse.

Try to use well-balanced external antennas on your transceiver; handi-talkie mounted rubber-duckie antennas are the worst, as they are inherently unbalanced and often create large unbalanced RF currents on audio wiring connected to the handi-talkie.

MICROPHONE JACK PINOUT AS VIEWED
FROM THE EXTERIOR OF THE TRANSCEIVER

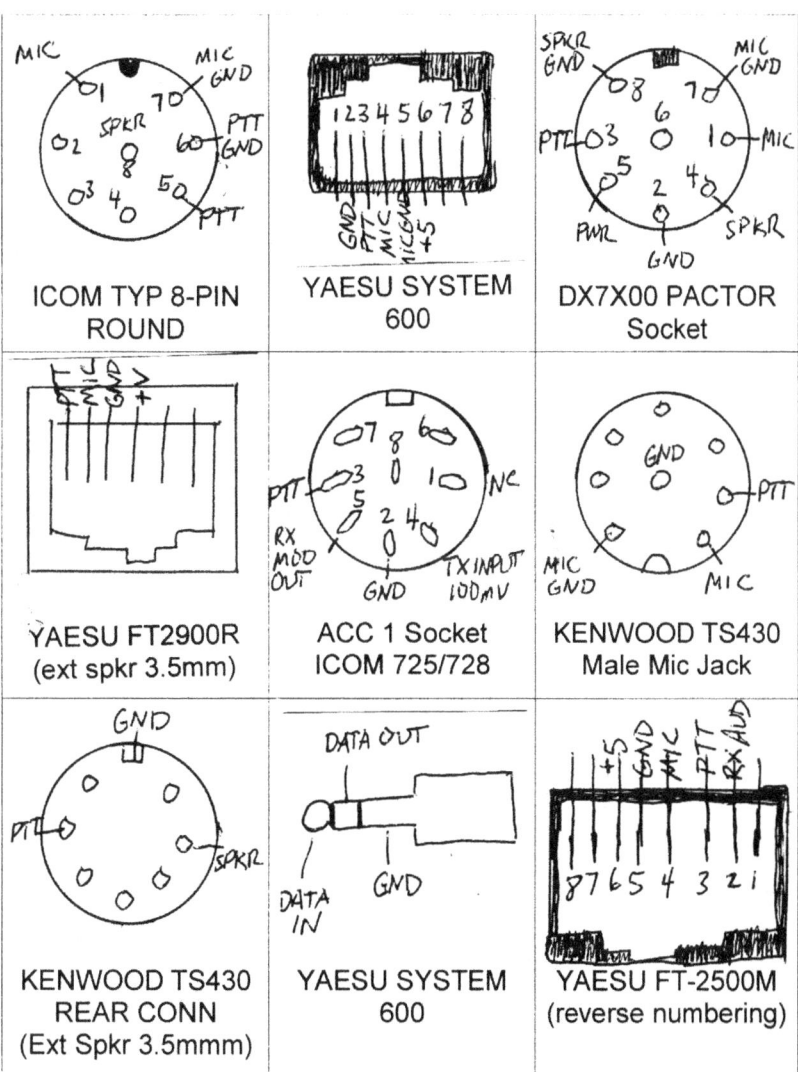

[DRAWING; G. GIBBY.]

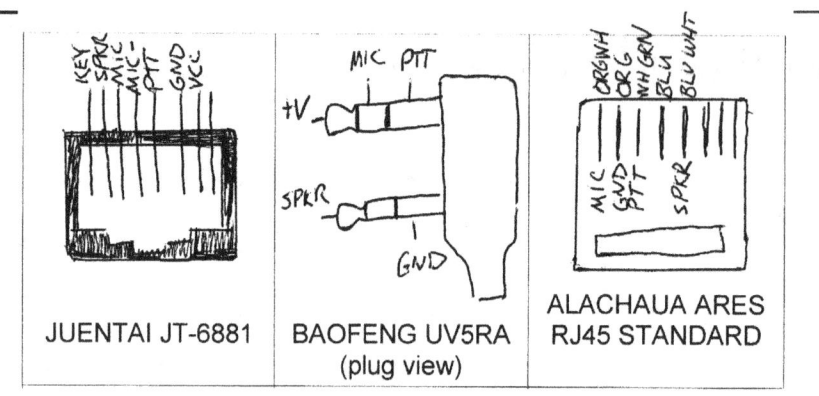

| JUENTAI JT-6881 | BAOFENG UV5RA (plug view) | ALACHAUA ARES RJ45 STANDARD |

DRAWING; G. GIBBY

12   REPEATER CONTROLLER -- ICS-CTRL

INTRODUCTION

The ICS-CTRL.com 1X (single repeater) raspberry-pi based repeater controller is a very inexpensive ($59, exclusive of the Raspberry pi) way to control a FM repeater, making the required identifications and even including the ability to make status announcements of voltages, temperatures and such, and respond to some commands.   I'm early in my knowledge of this product but it appears it can potentially provide Echo link connectivity also.   Since I have gotten comfortable with Raspberry Pi's (Particularly the Version 3 (model b)) for about $35 from Amazon, this is an attractive way to control an emergency (or any other) repeater.

However, although ICS-CTRL tries to make setup of their product duck-soup-simple, I still ran into some "gotchas" and a bit of knowledge of Raspberry and of linux goes a long way to get this product working. The vendor was extremely helpful, with email after email great suggestions and offers of telephone help if needed --- so I have compiled here all the information that I gathered in hopes of saving both later users and the owner the troubles that I had!

I believe that I know understand how to use a couple of soundcard USB-dongle type devices and some simple associated circuitry to use the openrepeater.com free software to duplicate the function of the ICS-CTRL product. His product is more I2C based and I had one raspberry that just didn't work with it --- but his product does work with the right Raspberry and it is a slick product once it is working!

Location of helpful software and documents:

Quick start guide to the 1X ICS-CTRL controller	
Manual for the 1X ICS-CTRL controller	
ICS-CTRL version of svxlink software for Raspberry Pi Model 3B	http://www.ics-ctrl.com/content/pi_images/ PI_REPEATER_2X_20180120_STRETCH.zip

	Note user **pi** password **ICS-CTRL**
ICS-CTRL version of svxlink software for Raspberry Pi Model 3B+ (introduced March 2018)	http://www.ics-ctrl.com/content/pi_images/ PI_REPEATER_2X_20180425_STRETCH.zip Note user **pi** password **ICS-CTRL**
Open Repeater version of software (note that it *does not appear to be able to control the I2C-bus-based ICS-CTRL controller* -- but it will work with USB-based sound cards	See: https://openrepeater.com/downloads Note: the open repeater group use the user root and the password is usually either OpenRepeater or openrepeater -- but it seems to be different from different access types. See this page: https://openrepeater.com/getting-started#5b80c3c50c3d5
Manual for the configuration file */etc/svxlink/svxlink.conf*	http://www.svxlink.org/doc/man/man5/ svxlink.conf.5.html
Video on hooking up cables	https://youtu.be/uqAOG9JomGs
Video on setting up input level	https://youtu.be/aSA_0I625Fw
Video on setting up output level	https://youtu.be/nGCcUR05pZU

HARDWARE

1.   Be very careful with the 1X (1X means able to control ONE repeater input receiver-output transmitter system) ICS-CTRL on the Raspberry Pi Model 3B that you don't let the tiny ICS-CTRL circuit board press down and short out against the conductive heatsinks below.   A bit of black electrical insulation tape might be in order here or a simple standoff system.

2. To properly insert the ribbon cable from the male radio pins on the ICS-CTRL board to the DB-9 female connector, you must know where pin 1 is on the male pins, and you must know that the BROWN wire of the ribbon cable is the #1 wire.   See photo below:

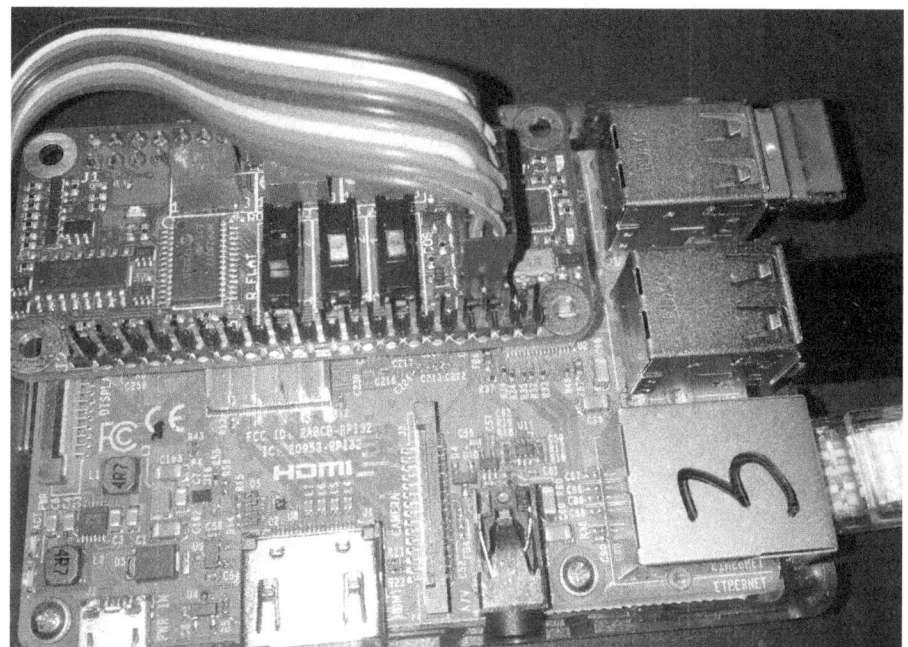

photo: G. Gibby

3. Until you are more familiar with the ICS-CTRL board, you might want to **take a fine tipped marker and mark the IN, OUT and GAIN potentiometers**.   The GAIN potentiometer controls the amplification of the output amplifier (going to the transmitter microphone input) and works in conjunction with the (linear) OUT potentiometer, which is in a normal volume control configuration.   This gives TREMENDOUS potential output variability in signal level!

The function of each potentiometers

Potentiometer (10 turn precision pot)	Clockwise Rotation Will
INPUT	Increase the sensitivity of the board to the signal from the receiver
OUTPUT	Increase the amplitude of the signal from

	the board to the transmitter
GAIN	DECREASE the gain (down to 1:1) of the output amplifier.

4.   The ICS-CTRL configuration of svxlink comes set for COR/COS sensing of receiver input signal, relying on a signal from the receiver that it has detected a valid signal.   This works well because even with the repeater user isn't yet speaking, the FM signal is constantly strong, constantly sending any CTCSS tones, and the receiver is easily able to detect the presence of the signal.   Most uhf/vhf FM radios (other than handhelds) include some sort of a signal detection output.   The ICS-CTRL board is able (by switch selection) to handle either active-high (high when signal detected) or active low. Handheld radios such as the Baofeng UV-5R don't come with such a signal.   There are some methods to create one from them, but for your initial testing (if you are using one of these) you may wish to simply have two wires from ground and the COS input (pin 7 of the DB-9) and short them to simulate this signal.

5.   DB-9 CONNECTOR PINOUT

        1- CTCSS ENCODE OUTPUT
        2- CTCSS INPUT FROM RX RADIO
        3- PUSH TO TALK (OPEN COLLECTOR) (OUTPUT)
        4- AUDIO OUTPUT (To Transmitter mic input)
        5- AUDIO INPUT   (From Receiver Audio output)
        6- GROUND
        7- COS INPUT FROM RX RADIO
        8- GROUND
        9- GROUND

SOFTWARE

1.   The ICS-CTRL software includes the i2cdetect program, which is incredibly helpful in detecting that your board's sound card has been detected.   See this page for more in depth information: https://linux.die.net/man/8/i2cdetect

Carefully note that the "2" is in between the i and the c.

```
i2cdetect 1
```

will cause your system (after a warning message) to scan the first range and you should see TWO "UU" entries in the resulting table, indicating that your board's soundcard has been detected.   (1X Model).

If you don't, something is awry and you should ask for help.   Note that this program can confuse things, so reboot after using.   (sudo shutdown now)

2.   LOCALIZATION: If you are accessing your ICS-CTRL board through the SSH protocol, you may not have any keyboard-layout issues.   However, if you are using a keyboard & monitor connected directly to the Raspberry Pi, you may have some characters/keys not properly working until you use raspi-config to set them --- and because of the vast array of choices, this can be problematic. The character set you may wish in the Unites States is: en_US.UTF-8

3.   Editing the `svxlink.conf` file. This is located in `/etc/svxlink/`   There are a multitude of text editors that can be employed.   You will likely need to use sudo vi or sudo nano or similar to be able to edit the file (which is read-only for user pi).   Be very careful obviously when editing this file.   As noted in the Table at the beginning of this information, the terse manual explaining all the options is found at: http://www.svxlink.org/doc/man/man5/svxlink.conf.5.html

4.   Identification: In the svxlink.conf configuration file, the times for "short" and "long" identification are given in MINUTES.   You may wish to shorten them temporarily in order to listen more carefully.   My callsign was not properly sent with the "/R" suffix until I added some SPACES after the callsign and forced their usage by placing quotation marks around the callsign in the configuration file. " NF4RC " worked for me.

5.   Receiver squelch issues: This is in the [Rx1] section of the ini-type configuration file svxlink.conf   There a a multitude of choices for receiver activation -- the ICS-CTRL software comes configured for

```
SQL_DET=GPIO
```

expecting the COS signal from the receiver.   Other choices are VOX, CTCSS, SERIAL, and several others (see the svxlink.conf manual cited above).   The VOX choice can be made to work marginally with uncommenting the FILTER depth and THRESHOLD options below it.

6.   AUDIO LEVELS & CONNECTIONS.   The ICS-CTRL system uses the well-known alsamixer to select the proper signals and set their gains.   You should probably run

sudo alsamixer

Then hit F5 to "show all settings" and be certain that CAPTURE is set to be "lin-in"
Then Set all sound levels to 60% and exit using the ESCAPE key.

I prefer to them store these settings using

sudo alsactl store

and you may wish to check them with sudo alsamixer if there are questions.

7. Getting the software to run -- it is supposed to be started as a background service in the boot up

sequence.   In my case I couldn't find it running (using `top`, **or**

```
top -d 0.25 | grep svxlink
```

The executable is located at

/usr/bin/svxlink

You can execute it manually by just typing that, but we want it to run forevermore....

Using my favorite editor I created the following file /home/pi/svxlinkscript:
```
# find svxlink process identifier
/usr/bin/pgrep svxlink
# if not  running
if [ $? -ne 0 ];
then /usr/bin/svxlink
fi
```

You can test that this file will start up svxlink by using

bash /home/pi/svxlinkscript

Then using the command **crontab -e** (and selecting the nano editor, which is self-explanatory) I appended two lines to the crontab file, and let it save it as it pleased:

```
# Each task to run has to be defined through a single line
# indicating with different fields when the task will be run
# and what command to run for the task
#
# To define the time you can provide concrete values for
# minute (m), hour (h), day of month (dom), month (mon),
# and day of week (dow) or use '*' in these fields (for 'any').#
# Notice that tasks will be started based on the cron's system
# daemon's notion of time and timezones.
#
# Output of the crontab jobs (including errors) is sent through
# email to the user the crontab file belongs to (unless redirected).
#
# For example, you can run a backup of all your user accounts
# at 5 a.m every week with:
# 0 5 * * 1 tar -zcf /var/backups/home.tgz /home/
#
# For more information see the manual pages of crontab(5) and cron(8)
#
# m h  dom mon dow    command
0 */12 * * * sudo /sbin/shutdown -r now >/dev/null
*/2 *   * * * bash /home/pi/svxlinkscript >/dev/null
```

The first line causes the raspberry to reboot every 12 hours and the second line causes it to check for svxlink every two minutes and if not running, it uses the script above to start it up.

13   HANDS ON SOLAR POWER

by Gordon Gibby KX4Z

Solar power set up ready to go -- 300 watt panel, battery, charge controller (on side of wood battery holder, which also has a 2 kW sine wave inverter.   Photo by G. Gibby

Let's work up a reasonable power budget:

Power consumed	Duty cycle	Power requirements
FM transceiver, 50watts output = 15A draw @ 13.8 VDC (	Assume 25% usage = 50 watts average	Over 24 hours, 1.2 kWhr

HF transceiver, 100 watt output, SSB and digital modes; 25A draw	Assume 15% usage = 50 watts average	Over 24 hours, 1.2 kWhr
Miscellaneous LED lights, laptop computer	25 watts	0.6 kWhr

Some conclusions:

Average wattage, radios and everything going	125 Watts
Assume we can re-orient the solar panels	10 hours of direct sunlight per day
Total 24 hour power required	3 kWhr
Assuming 10 hours of Sunlight per day	300 watts of solar power generation required
Amps of solar charging required @ 14 V	21 Amps
Batteries required to get through the night	125 Ahr worth of discharge capacity

Batteries

During the night hours, we'll be drawing this power from batteries -- and the nighttime usage will be about 12 hours * 125 watts = 1500 watt hours which is equivalent to 125 Ahr --- which means you'll best need TWO car-battery sized deep cycle batteries if you're going to run these radios all night.....or through a day with little sun!

This is a pretty heavy duty station, but there are 300 watt panels readily available. We might want to use two smaller panels.

Lets list some of the available systems -- much better choices than a few yeas ago: Because MPPT charge controllers have dropped significantly in price we can now make a very efficient system efficiently. MPPT is basically a computer controlled dc to dc converter that picks the exact best impedance to present to the solar panels (hence, what voltage to draw power from them) in order to get the most power into the load or batteries. The older Pulse Wave Modulation could only do a high speed "chop" of the solar panel output to avoid frying the batteries with an over voltage -- the MPPT literally can adjust the input voltage right to the best voltage to charge the batteries -- think of it as a very smart switching power supply and remember how little heat is given off by highly efficient switching ham radio supplies!

Solar Panels

Link	Price per watt	Power	Voltage	Weight	Price
https://www.wholesalesolar.com/1977280/astronergy/solar-panels/astronergy-chsm6610p-280-silver-poly-solar-panel	52 cents	280 W	31.2VDC	40.3 lbs	$145
https://www.wholesalesolar.com/1930020/canadian-solar/solar-panels/canadian-solar-cs6k-275p-silver-poly-solar-panel	55 cents	275 W	31 VDC	40 lbs	$150
https://www.wholesalesolar.com/9443231/solarland/solar-panels/solarland-slp090-24u-silver-poly-solar-panel	$2.25	90 W	24 VDC	17.6 lbs	$203
https://www.harborfreight.com/100-watt-solar-panel-kit-63585.html (includes pulse width modulation controller)	$2/watt	100W			$200

Charge Controllers

Link	Output 12V Amperage	Input Voltage (maximum)	Price
Victoreen BlueStart 75/15 MPPT https://www.amazon.com/Victron-BlueSolar-MPPT-Charge-Controller/dp/B00U3MK0CI	15 A	75 VDC	$88.40
Y-Solar https://www.amazon.com/dp/B078RJNDBM/ref=psdc_2236627011_t3_B00U3MK0CI	15 A	60 V	$50

120 VAC pure sine wave inverters

These are now available in much cheaper prices for 1, 2 and 3 kW -- expect to need a 2 kilowatt to start a refrigerator due to the high inrush current of the compressor.

Xantrex 2 kW (I have one and it starts a refrigerator) $367

https://www.amazon.com/gp/product/B002LGEMOQ/ref=oh_aui_search_detailpage?ie=UTF8&psc=1

14   EMERGENCY VHF ANTENNA BUILDING

This is one of the simpler tasks in ham radio, so don't let it frighten you!!! Most mobile operators using VERTICAL antennas on their cars, so the traditional polarization for FM vhf (or uhf) voice work is VERTICAL.

A simple dipole is just two wires connected to the antenna end of some coax. If you use bare wire, the total length from one tip to the other) will end up being near

    Length in feet =   468 divided by (Frequency in Megahertz)

    so for the center of the 2 meter band

    Length in feet =   468 / 146 = 3.2 feet = approximately 38 inches

    (each side will be about 19 inches)

Just strip the coax to get to the center and shield conductors, connect them to two bare wires (you can twist-tie, crimp, solder, it doesn't matter!) slap on some duct tape to keep them kinda held in place, and suspend your antenna with a string from the top end. (It doesn't matter whether the center conductor goes to the top or bottom ends.) The COAX needs to go out at RIGHT ANGLES so it is uniformly exposed to both the top and bottom fields from the antenna to minimize unbalanced currents, ideally for maybe 10 feet or so --- and then you can make a simple homebrew "common mode inductance" (aka "Balun") by adding in about 4 turns of the coax in a 4" coil, and add some tape or zip ties or string to keep it in the coil.

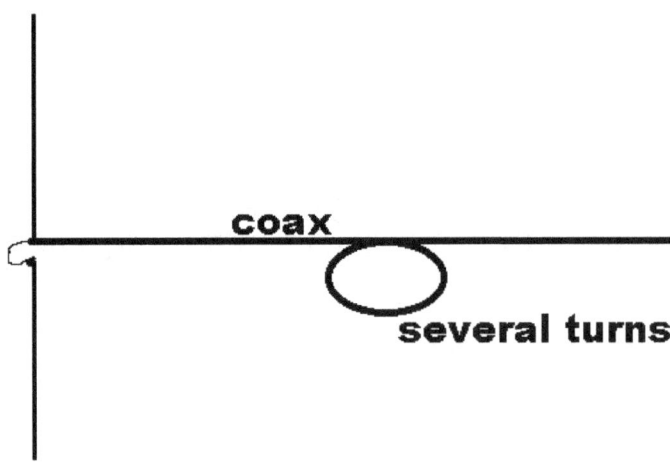

Drawing: G. Gibby

Couldn't be much easier!

If you use INSULATED WIRE, or if you affix your bare wire antenna to a yard stick or something for support, it will require a slightly shorter length by a few inches.   Taking an SWR meter with you will help for fine tuning but it isn't absolutely necessary as modern rigs are pretty well protected.

If you happen to have some FERRITES you can slip them onto the cable as well to reduce unbalanced currents --- none of that matters much if you are just doing voice work, but if there is a computer involved here....unbalanced currents can play havoc with Signalinks etc.

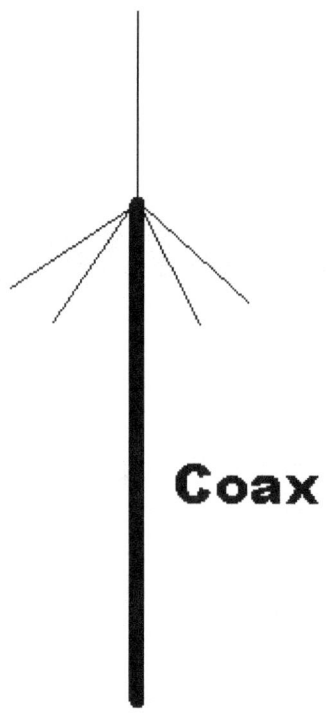

Drawing; G. Gibby

If you need to make an antenna fed with a VERTICAL coax run, then you can make a 1/4 wave vertical antenna with " 4 drooping Radials".   Run a vertical of about 19" (for the 2 meter band) upwards, and the add four evenly spaced 19" "radials"   at about 45 degrees from the vertical, and you'll reduce the induced unbalanced currents in the coax feed --- but you'll still probably wish to add the coil-balun or some ferrites to the line.

Pretty simple!!

15   TUNING A DUPLEXER WITH A $110 CHINESE SPECTRUM ANALYZER

by Gordon Gibby KX4Z

WHY

A "luggable" UHF repeater can be a huge asset for localized voice communications in an emergency when existing communications assets are damaged or overwhelmed. A repeater is a device that can simultaneously receive incoming radio signals and rebroadcast them (on a different frequency), often using a highly placed antenna that has significant reach to cover a wide area of client users whose antennas are at handheld or vehicle height. Point to point losses through foliage, vehicles and structures can be tens of dB per mile whereas free space losses from a client radio on a clear path above the trees to an emergency repeater atop the highest remaining structure, pulled up alongside a remaining tower, or on a tall geologic feature can be minuscule.

OVERVIEW OF HOW

The components of a repeater include a receiver, a transmitter and a controller that can accomplish the required periodic station identification. Although the repeater and transmitter are on different frequencies, no receiver is totally immune to strong off-desired-frequency signals, and a significant amount of "desensing" afflicts the receiver unless steps are taken to reduce the magnitude of the simultaneous transmitter reaching the receiver. This is the reason that table-top simple speaker-to-mic side-by-side dual handi-talkie "repeaters" are only good as long as you are basically in the same room with them -- in order to allow the receiver to come anywhere close to its normal sensitivity, you must protect it from that huge nearby transmitter that is the other half of the repeater! Vertical separation of two antennas by many feet can provide a significant degree of isolation as well as using large offset steps between receive and transmit frequency, but the commercial and amateur radio communities have settled on very high-Q electro-mechanical "duplexer" filters to keep transmitter and receiver isolated --- even when sharing the same coax to the same antenna! Typical duplexers provide upwards of 80dB reduction in the transmitter signal reaching the receiver. With a typical 50-60 dB off-channel rejection capability of good-quality receivers, the +40 dBm signal of the transmitter is thus reduced to -40 dBm at the receiver, allowing the receiver to achieve as much as -100 dBm sensitivity on the actual receive frequency, which is generally in the ballpark for useful performance.

This article describes an inexpensive route to putting together such a "luggable" portable UHF repeater. UHF is chosen over VHF because the physical size of the duplexer cans is so much more reduced by using the higher frequency band, where also typical receiver/transmitter offsets are 5 MHz (1.1% of frequency) instead of the 0.6 MHz of the 2-meter band (0.4% of frequency). The principles are the same for any frequency.

DISCLAIMER: Although I've managed to figure out a bit of this, and get a small system working tolerably well with very inexpensive measurement systems, I am in no way an "expert" on this – just a bit further along than when I started! There are so many different forms and versions of duplexer cavities.....I can only hope to address a small segment of these devices.   I'm providing links to some excellent articles about concepts that the reader would be well advised to consult.

DUPLEXER

The heart of the repeater is the duplexer.   New prices for quadruple 6" diameter duplexer can assemblies are past the $600 range. (Sometimes WAY past that price for well-made system.)   However, these are widely used for many services, including public service and GMRS and used systems can often be had at bargain prices, particularly at ham fests, because the real value of these systems comes only when they can be aligned properly and that is a rare skill.   I was able to get a used set for $50.   There are many different types of duplexer cavities, described as

- "band pass"
- "band reject"
- "notch"

and lots of combination descriptions and other descriptors.   The easiest to align seem to be those of the notch or band reject type, which typically employ a variable (glass-piston) capacitor in series with the injection loop.

Glass piston capacitors 4-15 pf
Mount in metal ground; solder to
far end. Photo. G. Gibby

All duplexer cans are going to have some means of mechanically adjusting the length (or effective length) of the 1/4 wavelength antenna "rod" firmly bonded to one end of the can --- to find a typical notch cavity, what you are looking for are duplexers where each can additionally has a screwdriver-adjustable (non conductive, please!) variable capacitor to set the "notch" offset from the resonant point of the duplexer enclosed rod.

Often these cans will have just one injection loop and either provide two BNC female connectors to it, or will use a coaxial adapter T-connector to provide two connections.   There isn't anything magical about can filters --- they just are a volume mechanical representation of a normal lumped tuned filter. For the simple notch or notch-band pass filter, we're first going to adjust for maximum PASS at one frequency, and then set the Notch (where the filter acts like a dead short across the conductors for the frequency that needs to be blocked.   Typically we can get 20 dB of notch or more from a single can. The coax between cans, and the subsequent can, can act like a voltage divider -- typically it is chosen to be near a 1/4 wavelength, so that it transforms the previous short circuit (at notch frequency) into a HIGH impedance, only to be shorted again by the next can (at notch frequency) furthering the notch.[1]

The duplexer cans between the transmitter and the antenna are set to
- PASS the transmitter frequency (headed to the antenna)
- STOP the receiver frequency – so they take out any spurious responses of the transmitter output.

The duplexer cans between the receiver and the transmitter are set oppositely --

- PASS the receiver frequency (from the antenna to the receiver)STOP the transmitter frequency – so the fundamental high power transmitter frequency doesn't make it to the receiver.

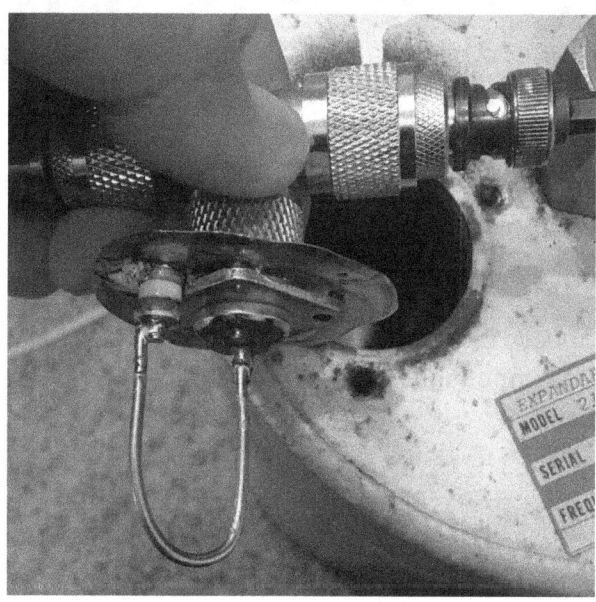

Homebrew illustrative notch loop with 12 gauge bare house wiring (inductor) and series glass piston capacitor (left side of loop) on simple metal circular piece that can be affixed above the portal in the can.   SO-239 and ground side of capacitor soldered to the metal base.   Photo G. Gibby

---

1   Because the injection loop may not present a perfectly resistive impedance, the optimal coaxial cable length to transform this into a high impedance presented to the next can may NOT be a perfect 1/4 wavelength.   Review the Smith Chart to better understand this.   Hence, some trial and experimentation are common.

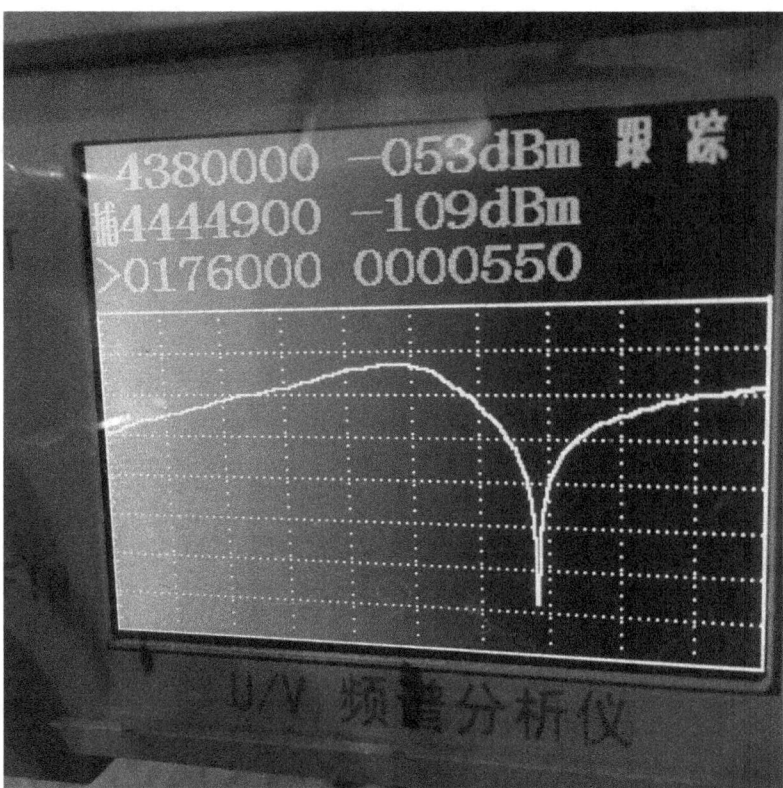

Excellent notch created by homebrew series tuned injection loop -- better than 50 dB, with a separation of 6.5 MHz in the 70 cm band -- used as illustration. Normal usage in this band is a 5 MHz separation
Photo G. Gibby

When tuning cans individually or in an arrangement, since these are impedance devices, **it is important that their source and load impedances are correct, typically 50 ohms.** A way to do this is to use attenuator pads, typically 6 dB pads, which tend to swamp out incorrect impedances and shift the source or load to appear to be 50 ohms. These can be purchased easily online (even on Amazon) but note that their power handling capabilities are very limited --- you'll pay extra to get one, for example, that can handle 2 watts. You'll need up to four of these. In order to handle the various tasks of tuning the duplexer cans, you'll also want 1 or 2 of the 20-dB attenuators. <u>One of your 6 dB pads should be rated at 2 or more watts.</u>

Double Shielded Coax
Its important that the cabling of a repeater doesn't allow bleed-through of unwanted signals around the high quality duplexor filter cans--- through incomplete shielding of the coax. Instead of typical RG58 or RG8, **one needs to use a double shielded cable such as RG400.**

You can purchase lengths of double shielded cable, but it probably to your advantage of get the correct crimper and learn how to crimp BNC and other connectors. Be aware that the RG400 shields are approximately RG8-size rather than RG58 size when you are purchasing crimpable connectors.

Cheap Ham Band Spectrum Analyzer

Using even a simple spectrum analyzer makes tuning the cans far easier.   A recently marketed simpleminded spectrum analyzer (basically a tunable receiver with a calibrated response that graphs its output versus frequency) with a tracking generator (so you can automatically generate a swept frequency and the spectrum analyzer reads out the amount of signal that is passed at each of the frequencies swept.

There are multiple tutorials online that give explanations for how to tune a duplexer system, first can by can, and then interconnected.

INITIAL TUNING

The inexpensive Chineese spectrum analyzer has a dynamic range in tacking generator mode from approximately -30 dBm to -80 or better dBm --- good enough to tune the individual cans easily and to start the process of tuning the complete array of the duplexer cans.   Mark each can with what it should pass and what it should block (assuming you have cans with a notch effect (caused by a series capacitor in the loop circuit).   Using pads on both sides, set the Chinese spectrum analyzer in tracking mode and adjust the tuning rod to get the pass frequency correct.   Then use the piston trimmer to set the notch in the right spot.   Repeat for each can.

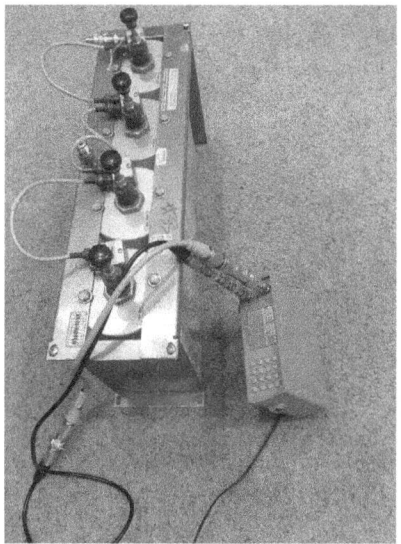

Photo showing pad attenuators on both input and output of inexpensive spectrum analyzer
Photo G. Gibby

When you connect the cans up into the full system, their tuning is likely to interact and change.   The exact length of coax between the cans matters also -- sources often suggest 1/4 wavelength (electrically -- you must correct for velocity factor) but in actual fact a slightly different length may work to give you better isolation due to inductive or capacitive effects of the injection loop in the can at the notch frequency.   So many sources make up an array of cables an inch or so off and try them all.....

Do your initial adjustment of the entire system using the Chinese spectrum analyzer in tracking mode – and don't forget to use 50 ohm termination on the "free" end – if you are adjusting the receiver side of the duplexer, you must put a 50 ohm load on the unused transmitter connection!

FINAL TWEAKS

However, to get the last bit of isolation, you need a signal stronger than -30 dBm so you can expand the effective dynamic range of your measurement out past 80 dB limit of the cheap Chinese spectrum analyzer. How? . The Chinese spectrum analyzer seems to get overloaded if you pass a signal stronger than -30 dBm through it (though it is supposed to be rated up to 0 dBm) --- so the solution is to forego the use of the tracking generator (that moves through a wide range of frequencies, at some of which the duplexer may pass a large signal to the spectrum analyzer). Instead, use the single frequency 1 watt output of a handheld transmitter, attenuated by a 6dB 2 watt-rated attenuator and then another 30 or more dB attenuation -- and **use the spectrum analyzer in non-tracking mode just as a plain spectrum analyzer**. The handheld now becomes your signal source and you use it to allow you to adjust *just the notch* to the best you can get it. Its dynamic range then extends all the way to -90 dBm and with a signal source attenuated to about 0 to -10 dBm you can measure the duplexer attention past 80 dBm

EXAMPLES:

Here is what ONE LEG of a typical GMRS repeater duplexer might look like before retuning to ham radio usage:

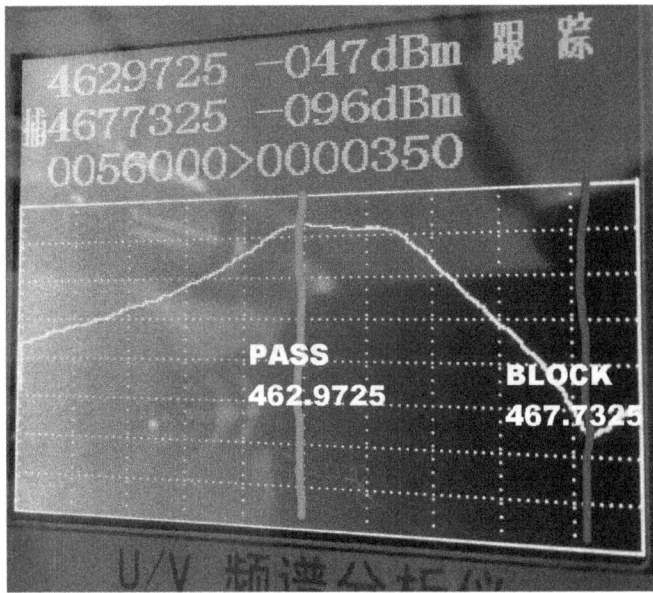

photo G. Gibby

Looking at the other leg

Photo G. Gibby

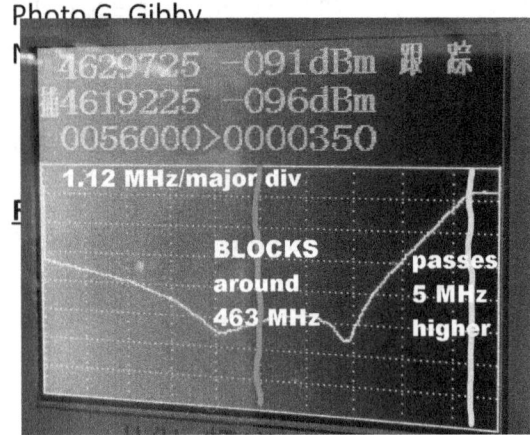

http://www.repeater-builder.com/tech-info/techindex.html	Long list of very useful papers for repeater building
http://www.repeater-builder.com/antenna/intro-to-duplexers/intro-to-duplexers.html	Introduction to duplexers
http://www.repeater-builder.com/antenna/pdf/emr-corp-antenna-duplexers.pdf	Introduction to duplexers
https://www.arrl.org/files/file/QEX_Next_Issue/Sep-Oct_2009/QEX_Sep-Oct_09_Feature.pdf	Relatively recent article o constructing your own cavities.
http://www.repeater-builder.com/projects/2m-duplexer.html	Older article on constructing your own cavities, and tuning with relatively primitive equipment

OPERATION OF THE CHINESE SPECTRUM ANALYZER

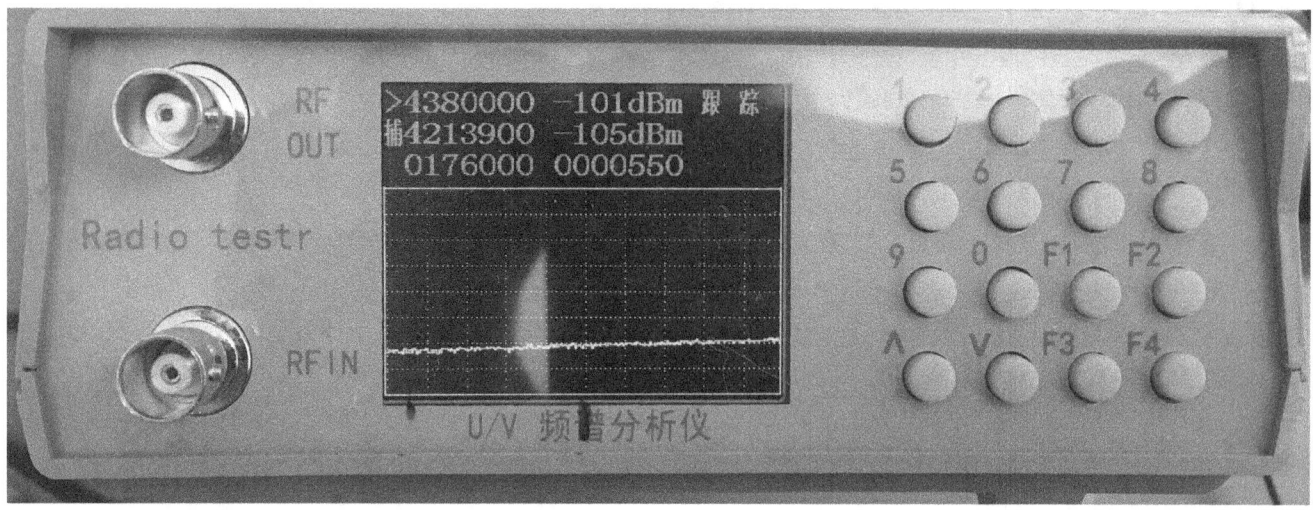

Inexpensive Spectrum Analyzer for the 2-meter and 70-cm bands.   Instrument is VERY small and lightweight -- note the size of the BNC connectors for comparison.   Photo by G. Gibby.

Theoretical Specifications
> - Frequency range: 136-173MHz 400-470MHz (seems to be that of the tracking generator)
> - Spectrum dynamic range: -127dBm --- 0dBm [optimistic?]
> - Resolution: 1dB
> - Tracking source output power: ≈ -38 dBm
> - Interface: BNC
> - Power: need external DC 8 --- 12V

Spectrum Analyzer Display

- Note that you cannot adjust the dB range of the screen -- the top is approximately -40 dBm
- Top of the display is -40 dBm
- Output of the source is approximately -25 dBm.
- 15 dB attenuation should bring it to about the top of the screen.
- Dynamic range of the system is approx 110 dBmmaximum   (from 0 to -110 dBm)

Numerical Display

- First Line: Center frequency of the display, and the power measured at that frequency. Frequency is in "tenths of kilohertz"
- Second line: Seems to display the frequency and power of either the highest or lowest signal detected.
- Third line shows the difference in frequency from the center to either right or left edge. The total range swept is twice this number.   Displayed in tenths of kilohertz.   Right hand number is the "step" of quantized frequency measurements.

BUTTONS
F1:

Short Press: cycle cursor between (a) center frequency (1st line), (b) frequency range (3rd line left), and (c) step cycle selection (3rd line right)
Long Press: switch between spectrum analyzer versus spectrum analyzer and tracking generator

F2: Backspace

F3 F4 Adjust the frequency in 1M

Increase / Decrease buttons: (up arrow, down arrow): the function seems backward to the USA user.

16 SOUNDCARD ISOLATOR

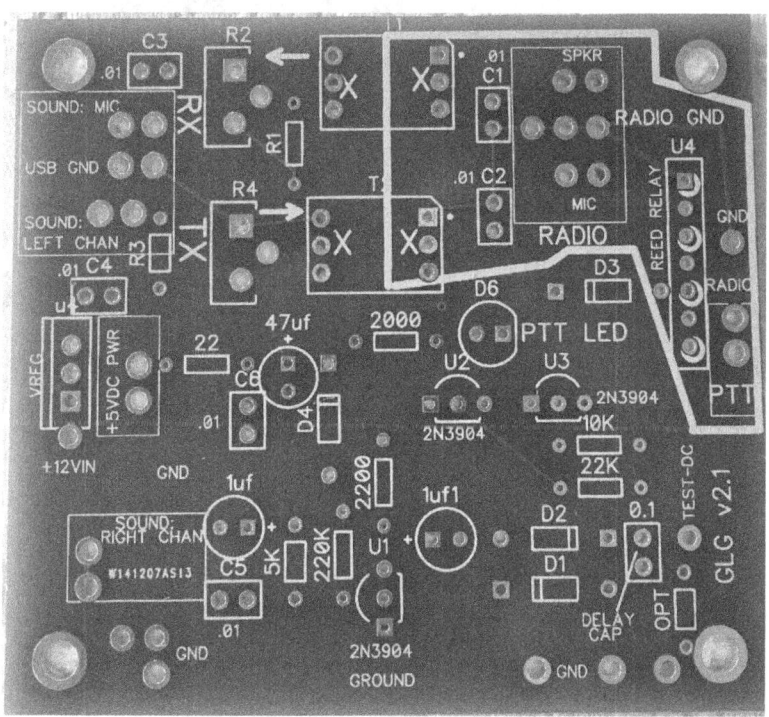

Printed Circuit board. Connections to the sound dongle/card are on the LEFT; connections to the amateur radio transceiver are on the RIGHT. Photos G. Gibby

By Gordon Gibby KX4Z

BUILDING INSTRUCTIONS

Use a low wattage soldering iron, such as 20-25 watts, with a fairly sharp and well-tinned (with solder) tip. While lead-free solder is encouraged today, 60/40 lead/tin rosin-core solder (still available) melts at a very low temperature, flows well and is very easy for beginners to work with. Try to avoid breathing the fumes!

The usual way to build a printed circuit board is to install a few components (3-5), bend their leads just a bit so that the component hugs the surface of the board, and then quickly solder each lead on the BOTTOM of the board with a well-applied iron and a quick touch of solder. **As soon as the solder melts and flow, remove both iron and solder to avoid over-temperature.** Solid state devices --- particularly transistors, can be damaged by too high a temperature for too long, so solder fast – only a very few seconds are required. Then clip off the excess leads (avoiding hitting an eye with the projectile lead). Move on to the next components. (Do not solder on the top of the board.)

CHECKOFFS (❑) are provided frequently in the following instructions to help you keep track of your progress and not skip any steps. These are optional.

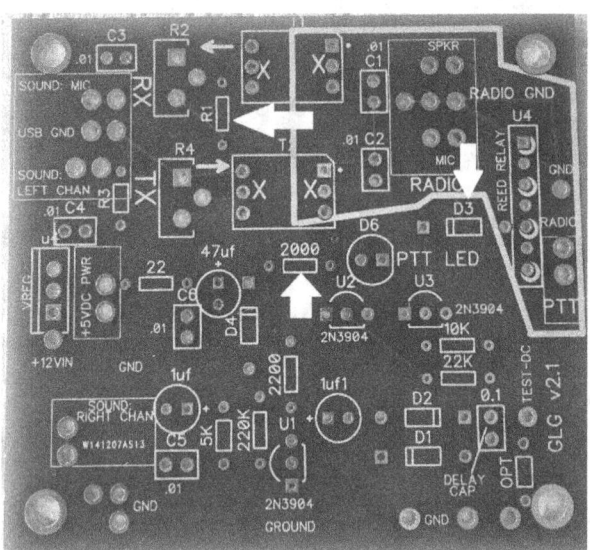

Components that are difficult to install once the transformer and relay are installed – and hence should go in first. Photos G. Gibby

The following quasi-step-by-step instructions are provided as a help (you don't have to follow them)

1. There are a few components that need to go in FIRST because other components make it impossible to add them later -- these are marked with big white arrows in the board photo above:

- ❑ R1 to the left of the top transformer (box with 6 holes, 2 of which are marked out with X's) -- use a 4700 or 5000 (5K) ohm resistor
- ❑ 2000 ohm resistor just below the bottom transformer
- ❑ D3 diode just to the left of the Relay (marked U4 when it should be RL1) (banded end goes to the left)

2. It is probably a good idea to do a few of the remaining components at a time, solder, and then snip their wires carefully.

3.   C1 and C2 are optional and usually we don't put them in. They were intended to quash radio frequency interference (RFI) from the radio and haven't been needed.

4.   Make ❑   R3 be 4700 or 5000 ohms. It isn't critical.   If you have too much gain, you can later reduce the gain by changing to 10K.

5.   All of the following capacitors are RF bypass capacitors and can be anything from 0.0047 uf to 0.01 uf

      ❑   C3
      ❑   C4
      ❑   C5
      ❑   C6

6.   ❑   The "delay capacitor" at the bottom right portion of the board must be 0.1 uf

7.   The following capacitors are electrolytic and may have either their (+) or (-) side marked -- and polarity is important and marked on the board

      ❑   47 uF just below and to the left of the lower transformer
      ❑   1 uF to the left of U1 2N3904 (+ side toward the transistor)
      ❑   1 uF to the right of U1 2N3904   (+ side toward the transistor)

8.   To avoid problems with radios that have a DC voltage present on their microphone line (to power an eletret microphone) ❑   **install a 1 uf capacitor with its negative side soldered to one of the two available "MIC" connections in the RADIO connections -- and later solder the wire going to your radio microphone to the unsupported positive end of the capacitor.**

9.   All of the diodes must be installed with their banded or notched end correctly as marked on the board:

      ❑   D1 near the bottom right of the board, banded end to the left-over
      ❑   D2 just above D1, with banded end to the right
      ❑   D3 to the left of the relay with banded end to the left (installed above)
      ❑   D4 below the bottom transformer with the banded end upwards
      ❑   D6 LED light – flattened end goes to the right-most

10.   Transistors have to go in so that their flattened and curved ends are correctly positioned over their screen printed drawing

      ❑   U1 2N3904
      ❑   U2 2N3904
      ❑   U3 2N3904

Note: While the 2N3904 is a nice transistor, it is likely that just about any npn small signal transistor would work.

11. ❑ Install the remainder of the resistors before mounting the
 ❑ two transformers (note the middle solder holes aren't used) and
 ❑ the Relay marked U4. Only the holes with white outlines are used for the relay, and the relay is built symmetrically and works no matter which end is "up"

12. Presuming you are going to get power for the board (5V at about 20 mA) from a USB source, you will apply power to the left side of the 22 ohm 1/4 watt resistor (used as a fuse) -- the +5V PWR solder holes can be used. If you are going to use a +12VDC input, you will need to add a 5 volt 3 terminal voltage regulator at the sockets on the left side of the board and feed it at the +12V input solder hole -- otherwise, the voltage regulator is not needed.

THINGS THAT WILL REMAIN EMPTY:

- ❑ 3-terminal voltage regulator on the left hand side unless you are going to power from 12V
- ❑ C1 and C2 bypass capacitors on the radio side of the transformers
- ❑ OPT capacitor spot near the lower right hand mounting hole – left in case you need to put in a switch to a higher delay capacitor to do CW.
- ❑ To give you "extra" solder connections, there are TWO solder holes for the sound microphone sound ground ("USB ground"), Sound left channel, +5VDC power in, Sound Right Channel, PTT output, Speaker connection to radio, Mic connection to radio, and 3 holes for the radio ground, as well as a spare radio ground hold near the Radio PTT. There are also six additional GROUND holes arrayed along the bottom of the board. These are optionally available to assist in making any connections you end up needing.

INITIAL SAFETY TESTING
Before applying power,

❑ very carefully look through the entire circuitry to be sure that you have put the right components in the right places,

❑ have polarized components (such as diodes, transistors, electrolytic capacitors, LEDs) inserted properly. This circuit has a "fuse" made of a 22 ohm "optional" 1/4 watt resistor to try and avoid damage to the USB bus of a laptop.

❑ When power is first applied, do it for only a second or two and watch for signs of untoward effects.

❑ It is normal for the LED to briefly flash when power is applied. Add a few seconds each successive connection, until the circuitry has proven itself safe.

ADDING USB +5VDC POWER

This is utilized merely to gain the +5V needed to operate the circuitry. Using any standard (type A) USB cable, open up the cable to gain access to the wires. These connectors are fairly standardized and the color code is usually:

USB CABLE COLOR CODE

Black = Ground
Red = +5V
White Data
Green Data

While avoiding shorting any of the colored wires to any other of the wires, use a voltmeter to verify the presence of +5V on the red wire while plugged into a USB receptacle.

With the proper wiring verified, connect the +5V to the left side of the 22 ohm "optional" resistor used for fusing purposes in this project. Connect the ground wire to the USB ground pads at the lower left portion of the printed circuit board.

❑ The collector of Q1's voltage (versus USB-ground) should be measured using a voltmeter. It should neither be saturated (0.2 V) nor cut-off (5 volts) -- it should best be somewhere in the middle (1-4 volts). If that voltage isn't right, something isn't correct with the biasing of Q1. Check:

a) right resistors in right spots?
b) solder connections all good?
c) +5 supply is really +5?
d) transistor wasn't fried during soldering?
....then consider getting help from a mentor.

WIRING THE SOUND DONGLE/CARD CONNECTIONS ON LEFT SIDE OF THE BOARD (COMING FROM COMPUTER)

The circuit requires connections to the headphone output and microphone input of a sound card, which can be the internal one of a modern laptop, or a USB-connected sound card dongle. The Adafruit 1475 inexpensive USB sound dongle is not the highest quality, but it is under $5 and works acceptably. Many other sound dongles will also work. One can open up some sound dongles and make direct soldered connections for the audio connections, but for this Conference, we will only discuss using stereo 3.5 mm (1/8") plugs.

STEREO AUDIO CABLE

First obtain two 3.5mm (1/8") stereo plugs with cables, preferably using normal wire, not the tiny nylon-embedded wire filaments with painted insulation used with some audio cables. Determine which wire represents ground, tip and ring of the jacks. The color codes in the table below apply to some batches of mass-produced cables – but check with an ohmmeter to confirm the proper wires of your kit.

Microphone Plug	Typical color code
Tip = microphone input	red
Ring = not used	white
Sleeve = ground	yellow

Headphone Plug	Typical color code
Tip = left channel	red
Ring = right channel	white
Sleeve = ground	yellow

❑ Wire the appropriate wires to the correct pads on the left hand side of the printed circuit board.

    ❑ Microphone input on the sound dongle
    ❑ Sound Dongle ground (which is usually the same as the USB ground)
    ❑ Headphone output left channels
    ❑ Headphone output right channel (lower left side of of the board)

You will probably want to mark the plugs so that you know which one to plug into mic and which to headphone output. Red tape or paint on the MIC plug is common.

The audio and USB cables can pick up RFI (radio frequency interference), and conduct it into the sound card, resulting in stuck-on transmitter, or locked-up computer or sound card.   Putting 2-3 coils of 2" diameter in these cables may reduce the pickup, as well as adding a ferrite bead around the cables.

    ❑ To test the PTT circuit, apply audio to the R CHAN in, with the computer or sound dongle output volume set to 100%, and the PTT LED should illuminate.
    ❑ BE CERTAIN THAT YOUR LAPTOP SPEAKER OUTPUT TO THE RIGHT CHANNEL IS AT 100% to properly activate the PTT LED & relay. You will adjust the level of the signal to be transmitted by the transmitter separately, using the TX (R4) trimmer potentiometer.

RADIO SIDE

The simplest method to connect to your transceiver is to solder four stranded wires from a multi-wire shielded cable to the mic / ptt / receiver audio / ground pads at the right hand side of the board, and then terminate the cable with the appropriate mic connector for your radio, in some cases needing also a 1/8" plug to obtain receiver audio output from an "external speaker" jack.   This method is reliable and works fine if you are going to use the sound card interface with only one radio.

If you are going to use shielded multi-pair twisted cable for this, **typically it is suggested to use the following colors for the radio signals:**

Signal	Wire Color
Mic	**White-Orange (remember this goes to the unsupported + end of a 1 uf capacitor soldered to the Radio Mic solder hole)**
Ground	**Orange (and bare shield wire)**
PTT	**White-Green**
Rx Audio	**White-Blue**

If you may utilize more than one radio, and need to be able to make connections to radios with varying connectors, you might prefer to provide an intermediate disconnection point (similar to how a Signalink provides a RJ45 socket). The popular Signalink accomplishes this with a female RJ45 jack on its rear panel. You can terminate the radio cable from the PCB with a male RJ45 and convert to a jack using a double-female RJ45 jack, which are less than 50c online[2] Then different cables, each with an RJ45 plug on one end, and the appropriate connector(s) on the other end for each radio, can be prepared for each different radio.

The question arises: what pins in the intermediate RJ45 shall be used for which signals? In the popular Signalink product, this is changeable for different radios to match commercially available radio cables. **However this introduces an unnecessary unknown (the pin out of the RJ45 jack) into equipment for emergency communications that would be best standardized.**

Our emcomm group in Alachua County standardized on one pin out, so that volunteers would always know the signal pin out on the sound interface card RJ45 plug, and as much as possible, we also use the same pin out on all Signalinks in use; it is the one recommended for popular Baofeng UV5RA radios, and also for mini-6-pin DIN radio connections.

ALACHUA COUNTY FLORIDA STANDARDIZED RJ45 PIN OUT

PIN	SIGNAL	Wire Color
1	microphone	white/orange
2	ground	orange
3	push to talk	white/green
4	unused	blue
5	receiver audio	white/blue

This table assumes the numbering shown in the accompanying Figure where the gold pins are held upwards and away from the reader.

Figure *Top View (pins visible) numbering of the RJ45 plug pins.*

---

2   https://www.monoprice.com/product?c_id=105&cp_id=10519&cs_id=1051902&p_id=7280&seq=1&format=2

This happens to be the standard pin out of commercially available cables intended to connect the popular Baofeng UV5RA (which uses a special molded double-plug connector) to a Signalink.   This allows one to easily plug in a Baofeng UV5RA low power transceiver to test a system.

You can also purchase surface mountable 8-pin RJ45 jacks which could be mounted on your enclosure. My local Home Depot carries those.

One inexpensive mounting, set to snap into a Twinings Tea Tin metal box. Photo G. Gibby

RFI REDUCTION

(A bit of trial and error, usually, and VERY important.)

To reduce RFI, put 3 or 4 loops (2-3" dia.) in the audio cable from the radio to the interface circuit board, and also clip on a ferrite "bead" if possible.

You may also put 3 or 4 loops (2-3" dia.) in audio cables to the USB sound card dongle if it is external to your metal enclosure.

If you used the direct-solder-to-sound-card technique and thus use a USB-extension cable from an internally mounted sound card dongle to your laptop, also put loops in that cable, and a ferrite "bead".

Laptop touchpads may frequently experience RFI when you touch them while transmitting. In that case, use a wireless mouse.

Try to use well-balanced external antennas on your transceiver; handi-talkie mounted rubber-duckie antennas are the worst, as they are inherently unbalanced and often create large unbalanced RF currents on audio wiring connected to the handi-talkie.

17   CREATING WINLINK ACCOUNT BY RADIO WITHOUT INTERNET

by Gordon Gibby KX4Z

NOTE: These instructions graciously provided by MIKE BURTON XE2-N6KZB, and incredibly prolific member of the Winlink Development Team who is all about helping people with their problems with WINLINK.   He once spent an hour on the phone with me to explain some settings to me.

1. Settings | Winlink Express Setup – put in your call sign, make up a password, be sure to put in a password recovery email because you'll want it when the internet returns -- fill in as much as you can in the remainder of the page.   You need to know your grid square but make one up if you don't

From WINLINK Google Group

2. When you try to UPDATE (to get out of the setup page) you'll get this message;

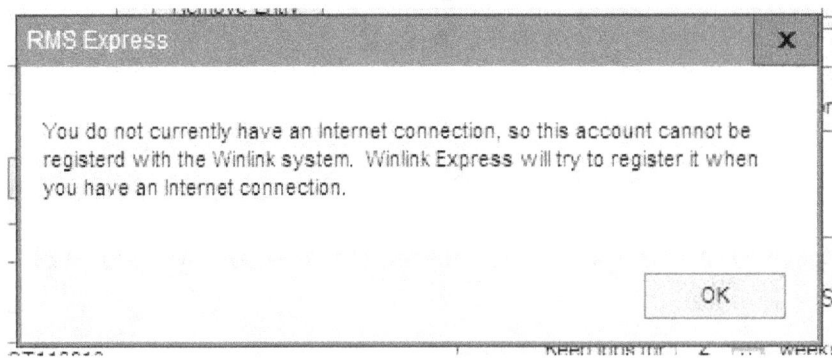

RMS Express

You do not currently have an Internet connection, so this account cannot be registerd with the Winlink system. Winlink Express will try to register it when you have an Internet connection.

OK

From Winlink Google Group

3. Now go make a WINLINK connection (pick any mode) – it will let you ONCE -- so pick a good station to connect to if possible -- and the system will send you an email like this one:

```
Message ID:  XE9FFFP7RK
Date;  2018/12/06  14:19
From:  SERVICE
TO:    XE9FFF
Source:  SYSTEM
Downloaded-from:   RMS:XE2BNC
Subject:  Your New Winlink Account

A new Winlink account for 'XE9FFF' has been activated .  The next time
you connect to a Winlink server or Gateway, you will be required to
use account password    <no quotes> .

In Winlink Express you'll find the option for configuring your
password under "Winlink Express Setup" in the Files Menu......  (GLG
note:   correct location is Settings | Winlink Express Setup)

You can manage your Winlink account........(details about getting into
the internet site)

It is important that you establish a password recovery address as
well!   (further details)
.....
Thanks for using Winlink
```

4. Take their advice and change your password in your Winlink Express software (Settings | Winlink Express Setup) to the new one they created for you!!!

From Winlink Google Group

5. If you don't take their advice and thus don't change your password to the one they assigned you, you'll get this outcome on the NEXT attempt to connect:

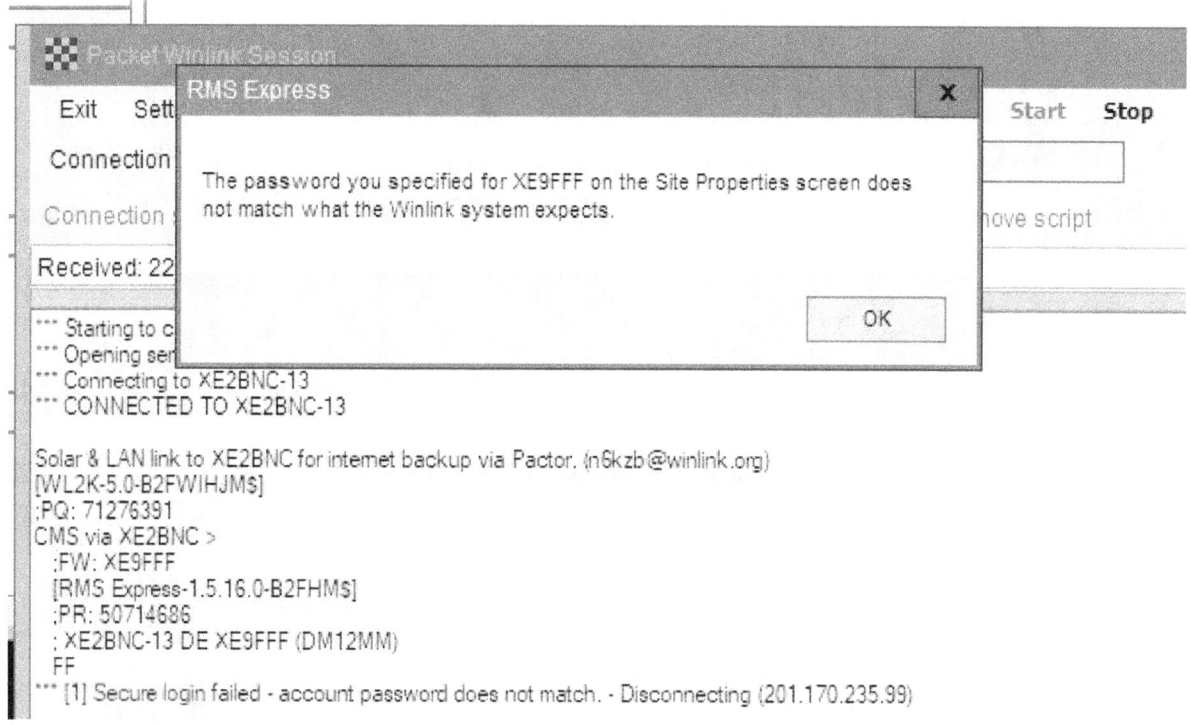

From Winlink Google Group

6. Once you get back to working internet – you can go to their web site (winlink.org) and change your password to whatever you wish by logging into your "account" there on their web site.

7. If you have never used WINLINK before, you'll probably also wish to download (over radio) the information needed about the various RMS Gateways ("channel information") and allow the system to create its predictive propagation information.

Many thanks to Mike Burton!

ABOUT THE NORTH FLORIDA AMATEUR RADIO CLUB

Created a couple years back in order to better support and advance emergency amateur radio communication systems in Alachua County, the NFARC club obtained the callsign NF4RC and maintains the web site

https://www.qsl.net/nf4rc/

through the graciousness of qsl.net.

This is the second emergency communications conference that our club has held. We've also held several "full scale" radio exercises, not only within our community, but even going so far as to deploy to Steinhatchee, Florida, site of a disastrous flood the year before our deployment.

Some of the texts our members have authored:

Amateur Radio Digital and Voice Emergency Communications
https://www.amazon.com/Amateur-Radio-Digital-Emergency-Communications/dp/1548004340

2018 Emergency Communications
https://www.amazon.com/Amateur-Radio-Emergency-Communications-Symposium-ebook/dp/B079JRYHHV

"The Blank Book"
https://www.amazon.com/Alachua-County-Emergency-Communications-Reference/dp/1724447084

"Steinhatchee Storm"
https://www.amazon.com/Steinhatchee-Storm-How-Puerto-Rico-volunteer/dp/1978441509

"Hurricane Irma"
https://www.amazon.com/Hurricane-Irma-After-Action-Report/dp/197773362X